THE
CLIMATE
FILES

THE
CLIMATE
FILES

THE BATTLE FOR
THE TRUTH ABOUT
GLOBAL WARMING

FRED PEARCE

guardianbooks

Published by Guardian Books 2010

2 4 6 8 10 9 7 5 3 1

First published in Great Britain in 2010 by
Guardian Books
Kings Place, 90 York Way
London N1 9GU

www.guardianbooks.co.uk

A CIP catalogue record for this book
is available from the British Library

ISBN 978-0-85265-229-9

Designed and set by seagulls.net

Printed and bound in the UK by CPI Mackays, Chatham, Kent

CONTENTS

CAST OF CHARACTERS

Main Players

Phil Jones has spent his entire career at the Climatic Research Unit (CRU) of the University of East Anglia, most of it in charge of assembling the past 160 years of global temperature records from around the world. Appointed sole director of the unit in 2004. Temporarily stepped aside in the wake of the publication of 13 years of the unit's emails. "I know I'm on the right side and honest, but I seem to be telling myself this more often recently."

Steve McIntyre Mathematician, former Canadian minerals prospector and, since 2003, full-time amateur scourge of climate science through his ClimateAudit website. "CRU's policies of obstructing critical articles in the peer-reviewed literature and withholding data from critics have unfortunately placed issues into play that might otherwise have been settled long ago."

Mike Mann Physicist turned paleoclimatologist. Moved from University of Virginia to Penn State University in 2004. Chief creator of the controversial "hockey stick" graph showing global temperatures over the past 1,000 years, based on a range of proxies such as tree rings. Co-founder of the RealClimate website. "This crowd of charlatans... look for one little thing they can say is wrong, and thus generalise that the science is entirely compromised."

Also Featuring

Raymond Bradley Climatologist at the University of Massachusetts Amherst. Co-author with Mann and Malcolm Hughes, a British-born tree ring specialist at the University of Arizona, of key early papers on the hockey stick. "As for thinking that it is 'better that nothing appear, than something unacceptable to us'... as though we are the gatekeepers of all that is acceptable in the world of paleoclimatology [this] seems amazingly arrogant."

Keith Briffa Tree ring researcher since 1977 at CRU, where he is currently deputy director. Helped develop the Yamal tree ring data sets. Early clashes with Mann mellowed into a long collaboration. Recently suffering serious illness. "I know there is pressure to present a nice tidy story... but in reality the situation is not quite so simple."

Judy Curry Climate scientist, Georgia Institute of Technology. Caught up in rows about climate change and hurricanes after Katrina in 2005. Subsequently renounced "tribalism" in climate science and sought common ground with critics. "Climate tribes were established in response to the politically motivated climate disinformation machine... to circle the wagons and point the guns outward in an attempt to discredit misinformation."

Mike Hulme Former researcher at CRU and founding director of the Tyndall Centre for Climate Change Research at the University of East Anglia. Interested in how climate change is represented in culture, history and the media. Recent author of *Why We Disagree about Climate Change*. "It is possible that some areas of climate science have become sclerotic… too partisan, too centralised. The tribalism that some of the leaked emails display is something more usually associated with… primitive cultures."

Doug Keenan A Canadian former financial trader on Wall Street and in the City of London. More recently, pursued a series of disputes with climate scientists, mostly about tree ring data. Accused Phil Jones and Wei-Chyung Wang of "fraud" for their handling of data from weather station locations in China. "It is unacceptable that the scientist who disseminates a data product on which international treaties are based… should actively seek to suppress information that calls the quality of the data into question."

Hubert Lamb Father of modern climatology and charismatic founder of CRU in 1972. Developed early theories about a medieval warm period and little ice age, and proposed the world was on the verge of a new ice age. Recruited Jones, Wigley and others, but later criticised what he saw as their simplistic theories of man-made climate change. "Increasing the carbon dioxide in the Earth's atmosphere – the end product of all our burning of fossil fuels and of vegetation – should raise the world's temperature, although the effect may be smaller, perhaps very much smaller, than is usually supposed."

Lord Lawson Senior Conservative politician and Chancellor of the Exchequer between 1983 and 1989. Recent campaigner against mainstream arguments on climate change, describing them as scientifically dubious and economically ruinous. Founded the

Global Warming Policy Foundation in late 2009. "Proper scientists, scientists with integrity, wish to reveal their data and all their methods. They do not require freedom of information requests."

Ross McKitrick Environmental economist at the University of Guelph, Canada. Senior fellow at the Fraser Institute. Collaborator with Michaels and McIntyre. "The key ingredient in most of the studies that have been invoked to support the hockey stick, namely the Briffa Yamal series, depends on the influence of a woefully thin subsample of trees and the exclusion of readily available data for the same area."

Pat Michaels Climatologist at the University of Virginia. Fellow at right-wing thinktanks the Cato Institute and George C Marshall Institute. Virginia State Climatologist. Recipient of funds from fossil fuel companies for projects such as his *World Climate Report*, published since 1994, and his "advocacy science consulting firm", New Hope Environmental Services. Rhetorical scourge of mainstream climate scientists, especially Santer and Mann. Author of books such as *Meltdown: the predictable distortion of global warming by scientists, politicians and the media*. "The truth is that what we sceptics say is always pilloried by the climate modellers, and then adopted as their own five years later."

Steven Mosher Statistician. Former "threat analyst" for Northrop Aircraft and writer of open-source software. After becoming unemployed in June 2009, he began analysing CRU temperature data in collaboration with McIntyre. Describes himself as a "lukewarmist". One of the first recipients of the hacked emails, but has never revealed who his source was. Published first instant book of the affair (*Climategate: The Crutape Letters*). "I called people mentioned in the mails. I read them mails… He passed me no information, just told me what I needed to know."

Benny Peiser Director of the Global Warming Policy Foundation. Social anthropologist, part-time lecturer at the School of Sport and Exercise Sciences at Liverpool's John Moores University. Co-editor of the journal *Energy and Environment*, regarded by Jones as "the worst journal in the world". "Personally I don't think the [email] disclosures make a big change to the overall scientific debate, but people are disillusioned… It's tarnished the image of British science round the world."

Jerry Ravetz Veteran US-born science philosopher at the University of Oxford. Studied scientific uncertainty. Proponent of the idea of "post-normal science" – a means of coming up with answers "where facts are uncertain, values in dispute, stakes high and decisions urgent", which he has lately applied to the problems of climate science after Climategate. "In retrospect, it is clear that the [CRU scientists] were practising evangelical science. For them there was a simple truth that would save us, and all naysayers were evil."

Sir Muir Russell Former head of the Scottish civil service and vice-chancellor of the University of Scotland. Appointed by the University of East Anglia to chair inquiry into Climategate. Most famous north of the border for his stewardship of the building project for the new Holyrood parliament building, which went massively over budget. "There are some very straightforward integrity issues to resolve."

Ben Santer Alumnus of CRU. Climate modeller at the Lawrence Livermore National Laboratory in California. Long-time adversary

of sceptic Pat Michaels after attributing climate change to a
discernible human influence in 1996 IPCC report. "[McIntyre]
has no interest in rational scientific discourse... He is the self-
appointed Joe McCarthy of climate science."

Gavin Schmidt Climate scientist at NASA's Goddard Institute
for Space Studies. Runs the RealClimate website, where he discovered the hacked emails had been uploaded in November 2009.
Has vigorously defended Mann, Jones and others caught up in
the Climategate saga. "Gravity isn't a useful theory because
Newton was a nice man."

Stephen Schneider Veteran climate modeller at Stanford University, California, and founder of *Climate Change* journal. Briefly
proposed in the 1970s that the world could face an ice age, before
becoming an early pin-up of climate change science. Prolific
author, most recently of *Science as a Contact Sport*. "We are
witnessing the 'contrarian Battle of the Bulge' now, and expect
that all weapons will be used."

Hans von Storch Meteorologist and director of the Institute for
Coastal Research in Geesthacht, Germany. "Soft" sceptic who
nonetheless resigned as editor-in-chief of the journal *Climate
Research* following publication of a sceptical paper he saw as method-
ologically flawed. "The authors of the damaging emails would be
wise to stand back from positions as reviewers and participants in
the IPCC process. The journals *Nature* and *Science* must review
their quality-control measures and selection criteria for papers."

Kevin Trenberth Climate scientist at the National Centre for
Atmospheric Research in Boulder, Colorado. Made enemies

among sceptics by linking hurricane intensity to climate change in the wake of Hurricane Katrina in 2005. "The fact is we can't account for the lack of warming, and it's a travesty that we can't."

Wei-Chyung Wang Climate researcher at the University at Albany in New York. Long-time collaborator with Jones on analysing Chinese temperature data. Accused by Keenan of scientific fraud over his contribution to a paper with Jones published in 1990. Exonerated by his university after an inquiry. "When we started on the paper we had all the station location details in order to identify our network, but we cannot find them any more."

Tom Wigley Australian-born climate scientist. Director of CRU from 1978 to 1993. More recently at the University Corporation for Atmospheric Research in Boulder, Colorado. Counsellor, critic and father figure in many of the emails. "Why, why, why did you... not simply say this right at the start?"

And

Myles Allen Head of the climatic dynamics group at the University of Oxford. Trenchant defender of emailing scientists in aftermath of Climategate.

Caspar Ammann Paleoclimatologist at National Centre for Atmospheric Research in Boulder, Colorado. Former student of Mann. Helped validate Mann's hockey stick with Eugene Wahl of Alfred University, New York.

Tim Barnett Marine physicist at the Scripps Institution of Oceanography, part of the University of California, San Diego. IPCC author. In 2005, published the first clear proof of man-made global warming in the world's oceans.

Sonja Boehmer-Christiansen Emeritus professor of geography at the University of Hull. Sceptic about acid rain and later climate change. Sees IPCC science as a "political construct". Avowedly political editor of the journal *Energy and Environment*.

Bert Bolin World-renowned Swedish climatologist who became the first chairman of the IPCC from 1988 to 1997. Died in 2007, shortly after the IPCC won the Nobel Peace Prize.

John Christy Climate scientist at University of Alabama, specialising in analysing satellite data. Spent 10 years as a Baptist minister in Kenya before taking up science. Mainstream researcher labelled a sceptic by some adversaries.

David Douglass Veteran astrophysicist at the University of Rochester and author of controversial paper attacking Santer's climate modelling analysis. Himself accused of statistical naivety.

Chris de Freitas Geographer at University of Auckland. Journal editor. Vocal climate sceptic and founder of Canadian climate sceptic body Friends of Science with his brother Tim, a geologist with Imperial Oil, ExxonMobil's Canadian subsidiary.

Jim Hansen Outspoken director of NASA's Goddard Institute for Space Studies. First scientist to get climate change onto front pages after warnings to US Senate during 1988 drought.

David Holland Electrical engineer from Northampton. Made a series of freedom of information requests to CRU, which, according to the Information Commissioner's office, were "not dealt with as they should have been under the legislation".

Warwick Hughes Freelance Australian scientist and among the earliest to question Jones's work on urban heat islands. In 2005, elicited from Jones the damaging question: "Why should I make the data available to you, when your aim is to try and find something wrong with it?"

George Monbiot *Guardian* columnist and environmental campaigner. Shocked many greens by calling for Jones to resign days after the publication of his emails.

Tim Osborn Climate modeller at CRU for the past 20 years. Member of the editorial board of *International Journal of Climatology*.

Lord Oxburgh Chair of the independent committee set up by University of East Anglia to examine the science done by CRU. Reported April 2010. Former chair of Shell UK and the House of Lords science and technology committee, current chair of a wind energy company.

Rajendra Pachauri Current chairman of the IPCC and director of Indian research agency the Energy Resources Institute. Tripped up in early 2010 for defending erroneous IPCC claims about the melting of the Himalayas – the affair known as Glaciergate.

Stefan Rahmstorf Oceanographer and climate scientist at Potsdam Institute for Climate Impact Research in Germany. Contributor to the RealClimate website.

Willie (Wei-Hock) Soon Harvard astrophysicist and senior scientist at George C Marshall Institute. Co-author with Sallie Baliunas of a controversial paper claiming new evidence of a medieval warm period.

Bob Watson British chemist involved in science policy-making as second chairman of the IPCC, head of environment at the World Bank and latterly chief scientist at Britain's environment department.

Anthony Watts Radio meteorologist, climate sceptic and founder of the world's most viewed climate website, Watts Up With That, which spread the word about the leaked emails. Says that despite his climate scepticism he is "green in many ways".

Zhaomei Zeng Researcher at the Institute of Atmospheric Physics, part of the Chinese Academy of Sciences, and visiting scientist at the University at Albany in New York, where she collaborated with Wang on collecting Chinese temperature data. Said she lost key location data on weather stations during a series of office moves.

Institutions

Climatic Research Unit (CRU) A research unit within the School of Environmental Sciences at the University of East Anglia in Norwich, England. Set up in 1972 by climatologist Hubert Lamb. Current director is Phil Jones. Previous directors include another prominent emailer, Tom Wigley. Other staff who feature in the narrative include Keith Briffa and Tim Osborn. The emails were copied from its server.

George C Marshall Institute First of the right-wing thinktanks in Washington DC to support sceptics working on climate change. Early supporter of Pat Michaels. Followed by the Cato Institute, the Competitive Enterprise Institute and others.

Goddard Institute for Space Studies Formerly the Goddard Space Flight Centre. Part of NASA, but a centre of climate modelling embedded within Columbia University in New York. Luminaries include its controversial director Jim Hansen and Gavin Schmidt, effectively the editor of the climate scientists' website RealClimate. Also early home of Stephen Schneider.

Intergovernmental Panel on Climate Change (IPCC) UN body set up in 1988 to publish periodic assessments of the state of climate science. Reports written by scientists but signed off by governments. Work on assessments completed in 1996, 2001 and 2007 is frequently discussed in the emails.

Met Office British government's meteorological service. Based in Exeter. Home to the Hadley Centre for climate research. Collaborated with CRU on compiling global temperature record for past 160 years. Hosted the IPCC working group, under chairman John Houghton, on climate science until 2001, when the National Oceanic and Atmospheric Administration took over.

National Centre for Atmospheric Research (NCAR) Hub for publicly funded climate research in the US. Based in Boulder, Colorado. Managed by the University Corporation for Atmospheric Research (UCAR), a consortium of US universities. Staff who feature in the narrative include Tom Wigley, Kevin Trenberth and Caspar Ammann.

National Oceanic and Atmospheric Administration (NOAA) US government agency heavily involved in climate research. Its campus in Boulder, Colorado, hosts the current senior staff and recent chair of the IPCC working group on climate science, Susan Solomon. Other staff in this narrative include Tom Karl, head of the US National Climate Data Centre in Ashville, North Carolina, and lead author on three IPCC climate science assessments.

Websites

ClimateAudit Created in 2005 by Steve McIntyre in Toronto, Canada, partly in response to attacks on his work by RealClimate, and as a vehicle for discussion of his audits of climate science. http://climateaudit.org

RealClimate Launched in December 2004 by Gavin Schmidt, Mike Mann, Stefan Rahmstorf and others as a home for commentaries by climate scientists "for interested public and journalists". Wide network of contributors. Part of the Guardian Environment Network. www.realclimate.org

Watts Up With That (WUWT). Created in 2006 by Anthony Watts, a California radio meteorologist, as a soapbox for mainly sceptical news and views on climate science. Perhaps the most visited climate website in the world. http://wattsupwiththat.com

Print Publications

Climate Research Published by Inter-Research, which was founded by Germany marine biologist Otto Kinne. He remains its publisher. Caught up in claims that it has been "captured" by climate sceptics.

Energy and Environment Minor journal published by Multi-Science, which aims to "fill gaps in the scientific literature". Has become a house journal for climate sceptics under its long-time editor Sonja Boehmer-Christiansen

Journal of Geophysical Research and **Geophysical Research Letters** Both run by the American Geophysical Union.

Nature One of two leading global English language peer-reviewed science journals. Based in London. Owned by Macmillan Publishing Group. Regular publisher of many of the emailing scientists.

New Scientist Popular weekly science magazine that has published this author's journalism for almost 30 years.

Science The second of the two leading global English language peer-reviewed science journals. Owned by the American Association for the Advancement of Science in the US.

INTRODUCTION
Windows on a closed world

"People said I should go and kill myself.
They said that they knew where I lived"

PHIL JONES IN THE WAKE OF CLIMATEGATE, MARCH 2010

Phil Jones entered the room, flanked by his vice-chancellor and by the man who had supplanted him as director of the Climatic Research Unit at the University of East Anglia while his conduct was investigated. His sleek grey hair and urgent stride to the microphone could not disguise his gaunt appearance and nervous demeanour. Jones had shunned the public spotlight during more than 20 years of investigations into the process of climate change. So this was a gruelling ordeal.

It was 101 days since his phone had started ringing at the university with demands from journalists for information about the alleged leaking of hundreds of emails sent and received by him and his colleagues over the previous 13 years. The reporters wanted explanations about the messages he had been sending. Apparently, he had been conspiring with colleagues as far away as Australia and California to subvert the normally sedate and scholarly discourse of science, and to silence a growing band of critics of his work compiling a record of more than a century of global temperatures.

Those 14 and a half weeks had been hell. The allegations about him on the internet had been vicious and frequently libellous. In their wake, his university had launched an inquiry into the work of his unit. His spiky American research collaborator, Mike Mann of Penn State University, faced a similar investigation over his emails to Jones and others. The British government's top official handling freedom of information legislation had said there was *prima facie* evidence that Jones had committed a criminal offence by suggesting data demanded by his critics should be deleted. And his vice-chancellor Edward Acton, while professing full confidence in Jones, was about to announce the appointment of a second review panel, to look into the scientific underpinnings of Jones's work.

In recent days, the university had allowed Jones to give media interviews. The 57-year-old revealed that, like several others caught up in the emails saga, he had received death threats. "I was shocked," he told one newspaper. "People said I should go and kill myself. They said that they knew where I lived. They were coming from all over the world." He had been on beta-blockers and sleeping pills, and had lost a stone in weight. He had contemplated suicide. He looked, friends and journalists alike said, a different man. Olive Heffernan began her article for *Nature*: "Phil Jones holds himself defensively, his arms crossed tightly in front of his chest as if shielding himself from attack."

Jones had made some minor admissions to journalists. His vast assemblages of data were sometimes "not well organised", he admitted. "Keeping track of everything is difficult." And yes, he might sometimes have suffered from a "bunker mentality" in repelling his critics.

Now he was out of the bunker and facing a grilling from MPs who were conducting their own inquiry into what had become known as Climategate. He was in the Grimond Room of Portcullis

House right by Westminster Bridge in London, just over the road from Big Ben. The escalators and glass-lined corridors of the new building were not like the fusty rooms and corridors of the Palace of Westminster. But the modernity could not hide the gravity of the interrogation. Even his friends had described it as, in effect, a court. "The outcome of these inquiries will potentially have as much impact on Phil Jones's professional standing and career as if they were criminal trials," Myles Allen, climate modeller at the University of Oxford, wrote to the *Guardian*.

Jones looked up at the TV cameras and the panel of MPs. He gulped. This was the first, and probably last, time he would face public questioning over Climategate. There was real sympathy in the room. Journalists like me, who had pursued his alleged misdemeanours with a puritanical vigour, paused too. Was it guilt? One parliamentary sketch-writer recalled a similar interrogation seven years before when another reclusive scientist, the UN weapons inspector for Iraq, David Kelly, had nervously faced cross-questioning from a parliamentary committee about weapons of mass destruction – and taken his own life a few days later. "The resemblance," wrote Simon Hoggart in the *Guardian*, "was disturbing and painful."

And it wasn't just journalists who were disturbed by what they saw. Jones's parliamentary performance was swiftly available around the world on YouTube. In the hours afterwards, a number of out-and-out climatic sceptics privately expressly their sympathy after viewing it. "I couldn't help feeling some sympathy for Phil, especially having known him in his fresh-faced-youth days and seeing him so visibly aged," one wrote in an email to me. "He is the fall guy," said another.

But any outsider unfamiliar with Jones's last 101 days would have been bemused. What could this diffident researcher have done to cause such conflicting emotions? If he was the fall guy, what had his colleagues been up to?

The Climategate story "broke" one Friday lunchtime in late November 2009. A huge file of emails and other documents from the University of East Anglia had been copied and published on the internet. They were red hot, my editor told me. The bloggers who first alerted the world to the presence of the liberated files – apparently on a Russian server in Tomsk – were already spinning them as a smoking gun that revealed a global conspiracy by scientists to dupe the world about man-made climate change. The bloggers, who mostly ran websites trashing mainstream climate science, had their quotes from the emails ready. And the interpretations. They knew that, as in most breaking news stories, whoever grabs the agenda first has a home run. Every subsequent story is refracted through those initial impressions.

I called Phil Jones. He wasn't a friend of mine, or even a regular contact. But I had known and reported on his work for years, usually when he published his annual estimates of the previous year's average global temperature. He was the keeper of the record of global warming. Jones already seemed to know what was coming down the road. He told me he thought Russians were behind it, because many of the emails concerned a recent row over climate data culled from tree rings in the Yamal peninsula in deepest Siberia. But whoever did the deed, he knew his critics were numerous and they would make mayhem.

I was one of the few journalists to speak to Jones before his university imposed a news blackout on its scientists' responding to the publication of the emails. The university did not want to discuss the content of the emails. So far as it was concerned, their publication was theft. There was a crime. The perpetrators had to

4

be found. Nobody, least of all journalists, should be dealing in stolen property. End of story.

But the emails were out there. Within a day, someone had put them on a website with a search facility. As I read the emails in the days after they were made public, it became clear to me that sniping over many years between mainstream climate researchers and their critics, many of them running climate-sceptic blogs, had degenerated into a secret and corrupting conflict. This was partly a rhetorical battle for the public agenda on climate change, but it went deeper than PR. It was also a battle for ownership of data, the building blocks of scientific theories, and a battle to open up the closed world of scientific peer review to challenge by outsiders. This was about more even than climate science; it was about the conduct of all science. A battle for the soul of science.

It became clear during the weeks after the leaking of the emails that their contents were also political dynamite, damaging to public confidence not just in the scientists themselves but in the probity of theories of man-made climate change. The highly selective quotes taken from the emails were having a real impact, made all the greater because the Climatic Research Unit (CRU) scientists were not responding to the allegations being made against them. Silence sounded like guilt. That was especially damaging because the quotes from the emails were being used in a way that was often hugely misleading.

Having covered many of the topics under discussion in the emails, I probably knew this better than most journalists. And as a bit-part player in the saga from my own appearance in the emails, I could see how my own words were being twisted.

One of the earliest emails in the file was from me to Keith Briffa, then a tree ring researcher at CRU and since promoted to be Jones's deputy. I had asked him to fact-check an article on his research I had written after a visit to CRU during early 1996. Fourteen years later, an American blog site called Baseball Media Watch had splashed this innocent exchange under the headline "'Biblical intensity' in search for sign of man-made global warming – Climategate email". The post included a couple of sentences from my draft article noting how, "For climatologists, the search for an irrefutable 'sign' of anthropogenic warming has assumed an almost Biblical intensity... The case remains 'not proven'."

Neither sentence made it into the published version of the feature – an edit that had nothing to do with Briffa, incidentally, only my editor's distaste for flowery prose. But in the fevered imaginings of the blogger, my draft became another whiff of cordite from the "smoking gun". This story was clearly going to run and run. And the mythology around what the emails said was set to grow.

It seemed to me the 1,000-plus emails deserved a thorough analysis. Treating them as a succession of soundbites was hopeless. They needed an investigation that explained the context in which they were written. I had met many of the scientists concerned and written about their work over the years. I had reported on their battles with the sceptics. Over Christmas 2009, I decided that I should try to write a short book about the affair.

The *Guardian* asked me to publish my findings at length, first in the newspaper and online, and then in book form. We hoped, by that process of phased publication, to gather more information and perspectives to complete the investigation. Some of what follows here is a result of a unique "conversation" with the *Guardian*'s readers, including key protagonists, who were invited to make annotated comments to the published online text. It was an exercise in open review.

Some people – scientists, climate activists and even some journalists – have argued passionately that writing this story is folly. They say that it can only create confusion and undermine public confidence in the basic, unchallenged science of climate change. I disagree. The emails are out there being misinterpreted and lied about on a scale that most of the besieged scientists simply do not understand. The only sensible response is to try to provide a more coherent and informed analysis.

The story also has a wider relevance. It raises important questions about how research is conducted; about whether the self-policing system of peer review is riddled with conflicts of interest, and how it should be reformed; and about whether, in an internet age, it makes sense to keep the world of science so closed. Might there be more to be gained from opening the process of scientific scrutiny to much greater public view? Later in the book, after examining the evidence revealed by the emails, I will explore these issues with the help of some eminent thinkers on climate science. They, and I, hope that something good can come out of this sometimes sorry affair.

PART ONE

Hotting up

CHAPTER 1
Two tribes go to war

"I'd really like to take a few of these 'auditors' in a dark alley"
BEN SANTER ON HIS CRITICS, APRIL 2007

This story is dark; there are few heroes. Neither mainstream climate scientists nor their critics come out of this very well. Some environmentalists will be distressed to discover what happens in the labs and how tea-room tittle-tattle seems to have become the preferred mode of communication among some respected scientists.

When the Climatic Research Unit's emails were subjected to public gaze they revealed that the climate scientists seemed traumatised by a long-term assault from critics they saw as politically motivated and commercially funded. The scientists believed their critics wanted to prevent them from doing their science and were intent on trashing their work. As their most vocal protagonist, Mike Mann, put it in an email to CRU scientist Tim Osborn in 2003, they wanted to hide from view "the sort of dirty laundry one doesn't want to fall into the hands of those who might potentially try to distort things".

It is striking that people with a limited scientific involvement with CRU who have been victims of past attacks – such as Kevin

Trenberth of the US government's National Centre for Atmospheric Research and Ben Santer of the Lawrence Livermore National Laboratory – became regular email correspondents with Phil Jones and his colleagues. Reading the emails gives the impression that they were huddling together in the storm.

Who were the critics? I think they were far more diverse than the scientists at the time acknowledged. Some fitted the stereotype of right-wing attack dogs – dinosaurs, as their opponents saw it, left over from the days before climate science had found its feet. But many did not. In particular, there was a new breed of critic without overt political or commercial motivation, amateur scientific sleuths driven more by curiosity and healthy scepticism for received wisdom. They wanted access to the complex data sets held by the mainstream scientists so they could test for themselves the latter's conclusions.

But the mainstream scientists came to hate and despise the sceptics, tarring them all with the same brush as wreckers intent on disrupting both their science and the wider global response to the dangers of climate change.

Faced with this response, even their least antagonistic and most thoughtful critics in turn came to believe that the scientists were not fallible human beings, but fraudsters. For these critics, climate science was being run by a handful of "gatekeepers" – a kind of scientific Politburo. They saw Jones, Briffa, Mann and other emailers reviewing papers by colleagues and rivals, passing or failing them like peremptory examiners. They saw how the gatekeepers also often held key writing positions within the Intergovernmental Panel on Climate Change (IPCC) in its assessments of the science of climate change, where they sometimes found themselves authoring assessments of their own work. So if they are damaged, then so is the IPCC. As climate contrarian Ross McKitrick put it in

October 2009: "The IPCC review process is nothing at all like what the public has been told. Conflicts of interest are endemic, critical evidence is systematically ignored and there are no effective checks and balances against bias or distortion."

The conflict between the two ideas about how science should be conducted – a closed system dominated by gatekeepers, or a more chaotic but less hierarchical open system – is the dominant story of the emails over more than a decade. The clash between the blogosphere and the pages of august journals such as *Nature* could not be greater. But within it is a Shakespearean tragedy of misunderstood motives. The bunker mentality – of climate scientists faced with the mob trying to take over their labs – is brutally exposed in the emails. But so too is the opportunism of the outsiders, and the confusion caused in the labs by their efforts to question what was going on.

The battle lines were not absolute. The emails reveal that some science insiders were demanding more openness from their colleagues, and providing candid criticism of shoddy or mistaken work. One person stands out in this: Tom Wigley. He was Jones's former boss, having preceded him as director of CRU. More recently based at the University Corporation for Atmospheric Research, in Boulder, Colorado, Wigley kept up a vigil for scientific honesty and integrity in emails over many years. But such individuals are rare. A number of people who sought common ground ended up ostracised and humiliated in the emails. And, as ever in war, truth was an early casualty.

Climategate would not have happened without one man: a Canadian squash-playing blogger and data obsessive in his early 60s called Steve McIntyre. Whether you see him as a hero or a villain, his data wars with Mann, Jones, Briffa and Santer helped create the siege mentality among the scientists, and set them on a path of opposition to freedom of information. By drawing in scores of data liberationists from both inside and outside the science community, he almost certainly inspired whoever stole and released the emails.

Though not an academic, McIntyre is a trained mathematician. He had a successful career heading small Canadian minerals companies, including being president and founder of Northwest Explorations until it was taken over by CGX Resources in 1998. He says he often used his statistical prowess to analyse mineral prospecting data and out-bet his rivals. In 2002, he took up a new hobby – investigating climate change science. He likes to say he began this task because he saw similarities between the deficiencies in climate science papers and the false prospectus that persuaded investors to buy into the world's most notorious mining fraud. Bre-X, an obscure Canadian mining company, rose to become one of the most valuable mining companies in the world on the back of what turned out to be lies about a massive gold find in the Borneo jungle.

McIntyre is balding and grey-bearded now, scruffy, personally diffident and with a slight speech defect. And, as these pages will show, he got up the noses of his targets through persistence and sarcasm. He has liked to give the impression that his minerals prospecting days are over, and that data is all he wants to mine these days. But he suddenly announced in March 2010 that his blog had "been a bit inactive" because he had been "doing some mining business in the past few weeks. One gold project, one zinc project." He was about to check in at the Prospectors and Developers Association of Canada convention in Toronto, his home town. Many climate scientists will be hoping that he has got the mining itch again, and will stop scratching them.

McIntyre investigated two areas of climate research. Both are dominated by large amounts of data analysed in some secrecy using methods that were rarely divulged and hard to replicate. One was the 160-year record of global thermometer readings compiled by Jones and his colleagues at CRU. The other was the creation of a history of temperature before thermometers, using various proxies such as tree rings. This came to be dominated by the work of Mike Mann, chief author of the famous 1,000-year graph of global temperatures known as the hockey stick. This used a range of proxies to show that global temperatures had barely changed over the past millennium until a big rise in the 20th century, creating the shape of an ice hockey stick on his graph.

McIntyre was a latecomer to the band of critics in both areas of research. Sceptics had been worrying away at Jones's data and Mann's hockey stick for years before he came on the scene in 2003. But his sheer persistence and evident obsession with verifying the data saw him explore areas never dreamed of by his more political forebears. And his online harrying unnerved researchers in a way that old-style rhetoric had not, as did his pioneering of the use of freedom of information legislation in the US and UK to demand the raw data behind the studies.

One of McIntyre's earliest efforts to probe the science of climate change was an email sent from his home to Jones at CRU in 2003 asking for some weather station data. Initially the exchanges were civilised. But as McIntyre's data demands grew greater, Jones became less accommodating and relations soured. Jones shared his concerns in emails, and the climate science community began to clam up when McIntyre called. The sniping became public. When Mann and colleagues started a website called RealClimate in 2004, to counter the claims of sceptics, McIntyre responded with his own site, ClimateAudit. Tit for tat.

Temperature records such as those published by Jones and his team at CRU are not simple records of past climate. They have been "manipulated" – a term that should not be regarded as pejorative, but does require openness about what is going on. As McIntyre's sometime co-author Ross McKitrick put it in his evidence to the House of Commons' Climategate inquiry: "Data sets published as climate records are not simply observations: they are the outputs of models that take weather records as inputs, apply adjustments aimed at removing non-climatic influences, group the resulting records into regional grids and then translate the data into deviations from local averages, yielding what are called gridded climate 'anomalies'."

The quality and integrity of the manipulation is vital to success. Even so, it has not been normal practice in the climate science community for scientists to publish either their raw data or the computer programmes they devise to analyse it. Only those in the know were aware of the methods in use, and even they regarded computer codes as private property.

McIntyre, an outsider, clearly doubted the analytical techniques being employed by the climatologists. He felt that, as a trained mathematician, he could do better. He certainly did not see his ignorance of climate science as a bar to this. And, as his requests for open dissemination of data and discussion of methods were rebuffed, he grew more suspicious. He questioned the reliability of the thermometer data used by Jones to produce his graphs of warming over the past 160 years. And he wondered exactly how Mann and others turned dozens of studies on the past climate, including a series of tree ring studies managed by Keith Briffa at CRU, into the neat hockey stick graph.

He found little evidence that scientists reviewing this work prior to publication examined the process in detail, nor any sign that independent researchers had seriously tried to replicate the findings. He concluded that the two cornerstone of scientific integrity – peer review and replication of findings – did not work effectively in the cases he was investigating. "Nobody's ever checked this stuff with any sort of due diligence," he said. Too much was taken on trust in the cosy, collegiate world of climate science.

Some mainstream climate scientists have come to agree. "One of the things that McIntyre and McKitrick pointed out was that a lot of the statistical methods used in our field are sloppy," says Judy Curry of the Georgia Institute of Technology, a climate researcher keen to find common ground with the sceptics. "McIntyre does not have a PhD. He does not have a university appointment. But he's made an important contribution, starting with the criticism of the hockey stick."

But that has been a minority view in the mainstream. Most climate scientists did not take kindly to this outsider invading their domain. They came to regard him as a meddling, time-wasting and probably politically motivated wrecker, who rarely published his own papers and devoted his retirement to undermining theirs. So when he tried to access their raw data and computer programmes, they resisted. The emails reveal that the researchers shared tactics, encouraged each other and competed for the rudest invective against McIntyre. "I looked at some of the stuff on the ClimateAudit website. I'd really like to take a few of these 'auditors' in a dark alley," mused Ben Santer one day.

And they grew even angrier as other wannabe investigators joined McIntyre's data hunt. Men such as Doug Keenan, a Canadian former financial trader on Wall Street and in the City of London, and a retired electrical engineer from Northampton called David Holland.

Some climate scientists have accused McIntyre, Keenan and others of being hired hands of corporations out to fight climate change legislation. The *Guardian* has found no evidence to support such allegations. Instead, many are numerate professionals with time on their hands, an obsessive streak in their heads and a cause to pursue. McIntyre says: "My interest has been entirely personal and has been carried out at considerable personal financial sacrifice in terms of opportunity cost in lost income... My wife is a teacher; our house is paid off; our children are grown up." Some, it should be said, also have a reckless disregard for civility. Keenan has accused several scientists of fraud and "criminal negligence".

The story of the battles fought by McIntyre and his acolytes to access the raw data, and the protracted and generally failed attempts by the scientists to repel him, is the central story of the leaked emails from 2003 onwards. As Santer, a persistent and vitriolic critic of McIntyre, has argued on RealClimate: "Mr McIntyre's unchecked extraordinary power is the real story of Climategate."

At first, McIntyre published regular peer-reviewed scientific papers, co-authoring a couple with McKitrick, a Canadian environmental economist. The mainstream climate scientists responded angrily to them. As we shall see, they often used their influence to exclude what they regarded as substandard papers from major journals. So McIntyre and other sceptical authors increasingly used *Climate Research* and *Energy and Environment* – two peer-reviewed journals at times widely disliked by mainstream climate scientists.

Tensions were strained further when McIntyre took to publishing more of his deconstructions of published papers on his website, without scientific peer review. Santer complained bitterly on RealClimate: "My research is subject to rigorous scrutiny. Mr

McIntyre's blogging is not. He can issue freedom of information requests at will. He is the master of his domain – the supreme unchallenged ruler of the ClimateAudit universe. He is not a climate scientist, but he has the power to single-handedly destroy the reputations of exceptional men and women who have devoted their entire careers to the pursuit of climate science."

In September 2009, RealClimate summed up how mainstream scientists felt about this kind of scientific discourse. "The timeline for these mini-blogstorms is always similar. An unverified accusation of malfeasance is based on nothing, and it is instantly telegraphed across the denial-o-sphere while being embellished along the way to apply to anything hockey stick-shaped and any and all scientists. The usual suspects become hysterical with glee that finally the 'hoax' has been revealed... After a while it is clear that no scientific edifice has collapsed and the search goes on... Net effect on lay people? Confusion. Net effect on science? Zip."

But, strident though his website often is, McIntyre has usually made a point of avoiding outright personal abuse. He never talks about a hoax being exposed, and has rarely questioned the central edifices of climate science. He just picks away at the data and computer codes, finding flaws, inconsistencies and sloppy checking. He provides fodder for his more excitable and less fastidious fans. And their abuse was usually only a link away on other sites.

Few of McIntyre's targets distinguished him from more politically motivated foes. For them it was a continuum. As the September 2009 RealClimate post went on: "Science is made up of people challenging assumptions and other people's results... What is objectionable is the conflation of technical criticism with unsupported, unjustified and unverified accusations of scientific mal-conduct." McIntyre rarely makes such charges personally but, the post complained, he "continues to take absolutely no responsibility for the ridiculous fantasies and exaggerations that his supporters broadcast".

Some have characterised all this as a simple clash of cultures, between the ways of Canadian mining prospectors and those of academia. As one academic put it to me: "I think McIntyre confuses the more aggressive and confrontational style of business he used as a geophysical consultant with the more even responses in scholarship exchanges." McIntyre dismisses this as "fantasy". They started it with RealClimate, he insists.

Whoever threw the first rhetorical punch, it is hard to argue, after reading the CRU emails, that the mainstream scientists are novices in abuse. Santer, for instance, concluded in one email to Tom Karl, head of the US National Climate Data Centre, in November 2008 that "McIntyre has no interest in improving our scientific understanding of the nature and causes of climate change. He has no interest in rational scientific discourse. He deals in the currency of threats and intimidation." McIntyre, he said, "is the self-appointed Joe McCarthy of climate science". When Australian climate sceptic John Daly died, Jones commented: "In an odd way this is cheering news."

In the final months before Climategate, the battle was not a cultural clash, or even really about climate change. It was about data pure and simple. McIntyre wanted the scientists' data – or good reasons why he could not have it. In one week in the summer of 2009, after being told CRU data was obtained from foreign governments only with confidentiality agreements, he and his ClimateAudit correspondents showered CRU with 58 freedom of information requests demanding to see those agreements – with, as we shall see, embarrassing results for Jones and his colleagues.

McIntyre often made it clear that he did not always have any particular scientific reason for requiring the data. He just wanted to liberate it. It was a battle to break down the walls of the ivory towers, to blow apart the cosy world of peer review. It was a battle for science, and for its lifeblood: data. Then came the stolen emails. Whether hacked from outside or leaked from inside, the emails lit a fuse, but the fuel of mistrust had been piling up for years.

CHAPTER 2

In the freezer

"The ice age is due any time"

GEORGE KULKA, CZECH CLIMATOLOGIST, 1971

The feuds over climate science go back a long way. The next two chapters will take us back to the beginnings of modern climate science.

In the mid-20th century, there was a chill across the world, and it wasn't just the cold war. From the 1940s to the mid-1970s, the planet seemed to be in the grip of a global cooling. For a while, almost every outbreak of extreme weather was blamed on it. Some members of a new scientific discipline, climatology, predicted a new ice age. Yet before the 1970s were out, temperatures were rising, and many of the same scientists were warning of global warming instead. It is a strange episode. Some of the personal stories that we will discuss later started back then. And some say it shows climate scientists are scaremongers and shouldn't be believed, whatever they are predicting.

Global cooling was a real phenomenon – and it changed history. In the winter of 1941, exceptionally cold weather stopped the German army's advance on Moscow: grease froze in German guns and thousands of soldiers died from the cold. Hitler's failure to take Moscow marked a turning point in the second world war. Without the freezing 40s, Hitler might have triumphed. But 30 years on no one was giving thanks for global cooling. As snow banks built up across the Canadian Arctic and pack ice grew in the North Atlantic, there was concern bordering on panic about where this might be leading.

In July 1971, Stephen Schneider, a young American climate researcher then at NASA's Goddard Space Flight Centre in New York, made headlines in the *New York Times* when he warned of a coming cooling that could "trigger an ice age". Soon after, George Kulka, a respected climatologist from the Czech Academy of Sciences, warned on TV in the US that "the ice age is due now any time". The US National Academy of Sciences reported "a finite probability that a serious worldwide cooling could befall the Earth within the next 100 years". As a hint of the horrors in store, weird weather in Africa led to a drought in the Sahel that starved millions. Today we tend to blame global warming for African droughts; back then, many blamed the cold.

Climate scientists called for action to halt the cooling. They included Fred Singer, first director of the US National Weather Satellite Service and today a well-known contrarian on global warming. Hubert Lamb, then in the process of establishing the Climatic Research Unit at the University of East Anglia, was of a similar view. The then editor of *New Scientist* magazine, Nigel Calder, adopted the same cause, making a TV programme called *The Weather Machine* that featured Lamb saying: "We should be preparing ourselves for a long period of mainly colder seasons... The little ice age lasted more than 300 years." And *Newsweek* gave a cover feature spot to an analysis of "the cooling world" written by science journalist, and sometime *New Scientist* freelance, Peter Gwynne.

Advisers to the Nixon administration in Washington DC proposed putting giant mirrors into orbit to direct more sunlight onto Earth. Australians proposed painting their coastline black to raise temperatures. Others suggested sprinkling Himalayan glaciers with soot to absorb heat and maintain the ice-melt that feeds the region's rivers.

What prompted this panic? Three decades of evident, if mild, cooling had set the scene. But there was also genuine concern among climate scientists based on predictions of both natural and man-made climate change. For one thing, the atmosphere was becoming dustier and filling with fine, light-scattering particles that were shading the planet's surface and, some suspected, causing the cooling. Reid Bryson of the University of Wisconsin-Madison argued that dust storms caused by farms spreading into more arid lands were mostly to blame. Others blamed urban smogs.

Schneider tried to calculate the likely cooling effect of this man-made air pollution. He compared it with the possible warming effect of carbon dioxide emissions, which it was by then clear were also accumulating in the atmosphere. In Schneider's early calculations, published in *Science* in 1971, the cooling effect was dominant. He said dust and sulphurous smog particles might have doubled since 1900 and could double again in the coming 50 years. Even allowing for warming from carbon dioxide, this could still mean a drop in global temperatures of 3.5C, which, "if sustained over a period of several years... is believed to be sufficient to trigger an ice age".

At the same time, research into the history and timing of past ice ages had found that there were many more than the four originally guessed at, their appearance driven by regular planetary wobbles. Worse, it was now clear that ice ages were the norm rather than the exception. According to Kulka, the most recent interval between ice ages appeared to have lasted only 5,000 years. Our present interglacial had already lasted 10,000 years. An ice age was long overdue. Perhaps pollution was already triggering its onset.

The early 1970s also saw the first analysis of Greenland ice cores, and with it the suggestion that climate could change very fast. The last ice age may have taken hold within as little as a century. So the cooling in the mid-20th century might not have been a short-term blip but the start of a rapid slide into the next global freeze.

Some climate scientists say today that the fad for cooling was a brief interlude propagated by a few renegade researchers, or even that the story is a myth invented by today's climate sceptics. Not so. There was, as Thomas Petersen of the National Oceanic and Atmospheric Administration (NOAA) showed in a detailed analysis in 2008, no consensus on global cooling. But equally, there was good science behind the fears. So why did Schneider and his fellow ice warriors get the prognosis so wrong?

One reason is that some of the calculations published with great fanfare were simply incorrect. Soon after his 1971 paper came out, Schneider realised he had greatly overestimated the future cooling effect from human-made aerosols. He had assumed that the increased concentrations of particles that he had measured in the air applied globally. They did not; they related only to small areas close to their source. Moreover, much of the particle load in the atmosphere turned out to be natural, so that even if emissions from human sources doubled in the coming 50 years, their effect would be much smaller than he had calculated.

Schneider also realised that he had underestimated the likely warming effect of carbon dioxide: it would be three times as great as he first calculated. When he redid the maths, the balance between warming and cooling now tipped strongly towards warming. In 1974, he published a retraction of his earlier prognosis – "just like honest scientists are supposed to do", he says today.

The science of ice ages has also advanced since. The planetary wobbles that periodically tip the world into ice ages are not identical, so some interglacial periods last longer than others. Good theoretical work now shows that the current interglacial is likely to be unusually long.

Most climate scientists now agree that the cold decades from the 1940s to 1970s had little to do with either man-made pollution or planetary wobbles. The mid-century cooling was mostly associated with two natural phenomena: first the eruption of a cluster of medium-sized volcanoes that pumped sunlight-scattering sulphate particles into the upper air, and second ocean oscillations such as the Pacific Decadal Oscillation, a kind of slow-motion El Niño that moved heat out of the atmosphere and into the oceans.

This does not mean that cooling particles have gone away since warming began. But, as Schneider eventually worked out, the accumulating greenhouse gases were of much greater importance as time went on. While the aerosols swiftly fell to Earth, the greenhouse gases hung around for decades or centuries. Most climate modellers agree that these aerosols are currently protecting us from some of the impact of greenhouse warming.

Celestial forces may one day have the final say on climate change, eclipsing any warming we have caused. Fred Singer, the great apostle of solar influences, may be proved right in the long run. But that is likely to be thousands of years away, and unless we cut our emissions of greenhouse gases soon, the argument goes, the next ice age might never happen at all. Tim Lenton of the Tyndall Centre for Climate Change Research, a neighbour of CRU at the University of East Anglia, recently calculated that global warming could reach 13C in the coming centuries, double the difference between today and the depths of the last ice age.

All this raises an alarming question. If climatologists were so wrong in the 1970s, why should we believe them now? One reason is that, as those who played a part in the cooling scare now readily admit, the early studies were based on flimsy data collected by very few, often young, researchers. In 1971, when Schneider's paper appeared, he was instantly regarded as a world expert. It was his first publication.

Today, far more research has been done into how and why climate changes. The consensus on warming has grown much bigger, much broader, much more sophisticated in its science and much longer-lasting than the spasm of concern about cooling. That doesn't make it right, of course. All science is provisional. Old, experienced researchers can be wrong, too – even the best and most celebrated. Take Hubert Lamb.

The Climatic Research Unit was established at the University of East Anglia in 1972 during a flurry of concern about global cooling. Its founder was one of the legends of climate science, Hubert Lamb. Its aim was not to forecast the future, but "to establish the past record of climate over as much of the world as possible, as far back in time as was feasible".

Lamb was one of the first meteorologists to argue that climate was not fixed, but constantly varied. He had two big and controversial ideas back then. One was that there was a marked warm period in the history of Europe's climate, and perhaps more widely round the world, centred on the 11th century. Lamb called it the medieval warm period. He originally got the idea from a study of evidence for temperatures in central England. But by the time he published in 1965, he was confident that "historical records, as well as archaeological, botanical and glaciological evidence in various parts of the world from the Arctic to New Zealand... have been found to suggest a warmer epoch lasting several centuries between about AD900 or 1000 and about 1200 or 1300." Why else would there have been vineyards in northern England or settlers farming the shores of southern Greenland?

He added that a subsequent "cold epoch" was now known to have "reached its culminating stages between 1550 and 1700". This was the little ice age.

His second big idea, as we have seen, was that the world was most likely headed towards a new ice age. He recognised the greenhouse effect of elevated levels of man-made greenhouse gases, which were first measured regularly from the late 1950s by the American scientist Charles Keeling. But he reckoned that those gases were unlikely to prevent the coming cooling. Lamb became known among British climatologists as the "ice man".

Lamb retired in 1978, to be replaced as director of CRU by the Australian climate modeller Tom Wigley, who dedicated the unit to pursuing evidence of climate change out of concern about the warming effect of greenhouse gases. That was also the year CRU began the creation of its global thermometer-measured temperature data set, which became Jones's life's work.

Starting in 1986, Wigley and Jones published regular global analyses and updates of world temperatures. These coincided with a period of global warming that began in the mid-1970s, with the new certainty that carbon dioxide was accumulating in the air, and, as we shall see in the next chapter, with a growing sense that these two trends were linked, and were of crucial importance for the planet's future. A new generation of scientists at CRU and elsewhere sought to demolish many of Lamb's ideas.

But in 1984, in what in retrospect seems like a last hurrah, Lamb gave a lecture to the British Association for the Advancement of Science on "the future of the Earth" that sounded a warning to their endeavours. On carbon dioxide, he said: "Increasing the carbon dioxide in the Earth's atmosphere – the end product of all our burning of fossil fuels and of vegetation – should raise the world's temperature, although the effect may be smaller, perhaps very much smaller, than is usually supposed." On past carbon levels in the atmosphere, he noted "sharp maxima... at or soon after AD1100, and almost as high again briefly in the early 1500s. The magnitude of the variations indicated before the industrial era is surprising and the reasons... need explaining." On current trends, he noted that "attempts to present the whole rise of world temperature since the Industrial Revolution... as attributable to the increase of carbon dioxide... [are] unwarranted."

He continued with a paternal warning to what he saw as a new generation of impetuous young researchers, many of them appointed by him at CRU: "The eager search in many laboratories for verification of the expected carbon dioxide-caused warming of the climate may have been pressed too hard. There is danger in the situation where scientific research is richly funded by interested parties." And finally: "There is unfortunately room for error and misjudgment in the preparation of temperature series which attempt to present averages and trends for the whole Earth... A large number of temperature-observing stations on land are increasingly affected by artificial sources of heat, particularly from urban growth." As we shall see, this could have been designed as a message to his young CRU protégé Phil Jones. Perhaps it was.

Lamb's pioneering 1965 paper on the medieval warm period was referenced in the first IPCC report in 1990, which included a diagram drawn by Lamb apparently showing the period as warmer than today. But it subsequently disappeared. And the idea of a new ice age has long since been consigned to the lab dustbin.

Lamb died in 1997, five months before the signing of the Kyoto Protocol on greenhouse gases apparently condemned some of his views to the history books. As CRU's Mick Kelly delicately put it in an obituary, the old man "was sceptical of certain claims regarding the dangers posed by global warming". The evidence has grown to suggest that he was wrong. But, though sceptical about the force of man-made climate change, he was no denier. He concluded his lecture to the British Association in 1994 with the words: "The knowledge that we now have as a result of climatic studies contributes strongly to the dawning appreciation of our responsibilities to care for the future of the environment which the Earth provides for mankind, for the animal kingdom and for the plant world."

Nonetheless, there is an irony in the fervour with which Lamb's successors have overturned his scientific traces. For the science discussed in the emails is mostly the work that he set in motion to "establish the past record of climate over as much of the world as possible, as far back in time as was feasible". It is done by assembling data sets from the readings of thermometers in weather stations round the world, from satellite measurements, and proxy measures of historical temperatures such as tree rings and ice cores.

The result has been iconic graphs like the famous hockey stick, first published 12 years ago and one of climate science's most famous and controversial products, showing a long period of natural stable temperatures followed by a sharp, exceptional warming in the late 20th century. It appears to refute Lamb's idea of a medieval warm period and a little ice age, or at least relegate them to phenomena for Europe and the North Atlantic, rather than a global force.

In this work, CRU has been crucial. It has assembled one of the most comprehensive thermometer data records in the world, much of it under contract to the US Department of Energy. And Jones has been central to that activity. He joined the unit in 1976, at the age of 24, when Lamb was still the director. He was a full-time researcher, engaged mostly in compiling the famous temperature data set, for almost two decades. His first paper on the topic appeared in 1982. Then he began to combine research with teaching duties, before becoming first joint director of CRU in 1998 and then sole director from 2004. His name has appeared on 270 peer-reviewed papers, roughly one for every six and a half weeks since he arrived at the unit.

CRU is also home to some leading tree ring researchers, notably the current deputy director, Keith Briffa, whose analysis has been central to creating the hockey stick. CRU's work is the prime (though not the only) basis for the claim that man-made global warming is happening now and is exceptional in history.

Lamb is not forgotten at CRU. His research findings may have been confined to the deep freeze, but he remains revered. In 2006, Jones presided as they renamed the unit's main building the Hubert Lamb Building. But as CRU came under siege from its critics in recent months, some of Lamb's past sceptical words about man-made climate change may have echoed through the CRU corridors. However hard they resisted, the suspicion must surely have surfaced in a few hearts that the old man could have been right all along.

CHAPTER 3

The secret origins of the IPCC

*"It was a catalytic event. Suddenly the climate
change issue became much more urgent"*
BILL CLARK ON THE 1985 VILLACH CONFERENCE

There were no journalists present, and no headlines afterwards. But
on 9 October 1985, at the sleepy town of Villach in the Austrian
Alps, 89 scientists from 23 countries gathered for the International
Conference on the Assessment of the Role of Carbon Dioxide and
of other Greenhouse Gases in Climate Variations and Associated
Impacts. When they left six days later, they had turned debate about
climate change from a backwater for a handful of atmospheric physicists into one of the great global issues for the decades ahead. Forget
global cooling; it was global warming the world had to worry about.

In a long concluding statement, they declared that "in the first
half of the next century, a rise in global mean temperatures could
occur which is greater than any in man's history". They warned for
the first time that the likelihood of this catastrophe happening
would be "profoundly affected by government policies". And they
called for the UN to "ensure periodic assessments of the state of
scientific understanding" of climate change along with "consideration of a global convention" to tackle the threat.

"Villach was the turning point," says Jill Jaeger, an environmental sciences graduate from the University of East Anglia who was dragooned into editing a report for the meeting. It was a "catalytic event", says Bill Clark of Harvard University, the author of key scientific studies on climate change in the US in the early 1980s. "Suddenly the climate change issue became much more urgent," Bert Bolin, the leading Swedish meteorologist who oversaw the meeting's scientific report and later became the first chair of the IPCC, told me before his death in 2007.

Everyone who took part agrees that the driving force was Jim Bruce, a leading Canadian climate scientist and then the number two at the World Meteorological Organisation (WMO). He chaired the Villach meeting. "It was Jim who wanted a statement from the meeting that ensured political follow-up, and he coordinated the group that drafted the statement from the meeting," says Jaeger. He also conceived of the idea of a new body to assess climate science. So if anyone laid down the tram lines for the IPCC, and if anyone introduced politics into the DNA of climate change science in the coming decades, it was Bruce.

There were others working behind the scenes. Along with the WMO, the meeting was co-sponsored by the UN Environment Programme. UNEP's influential director, the Egyptian scientist-cum-diplomat Mostafa Tolba, had earlier that year brokered a global treaty to protect the ozone layer – the world's first global environment treaty. He now wanted to repeat the trick with greenhouse gases.

But people who were there say the key to the meeting's future influence was the fact that the international group of scientists met in Villach as individuals, rather than as representatives of governments or scientific bodies. They could say what they thought. And did.

This was unusual for climate science at the time. Until then, with the exception of a few individuals like Hubert Lamb, who was not in Villach, the topic had been largely the preserve of government research bodies like those attached to the US Department of Energy. Caution had been the watchword in pronouncements on future climate change. Let off the leash, the researchers signed up to a forecast of "substantial warming" that would be unambiguously "attributable to human activities". No ifs, buts or maybes.

Wendy Franz, who wrote a report later on the evolution of climate policy for Harvard University, reckoned it was impossible to overestimate the effect of the absence of domestic political constraints. This was particularly true for the Americans at the meeting. They were pioneers of much of the basic climate modelling, but they were operating under the Republican administration of Ronald Reagan, which was hostile to any suggestion of constraints on burning fossil fuels.

That made American scientists ultra-cautious in pronouncements with any potential significance for policy. Just two years earlier in 1983, meeting under the umbrella of the US National Research Council (NRC), several of the American participants at Villach had written a report for their government on climate change science that stressed the "continuing scientific uncertainties", and the need for more research. But the Villach deliberations made them see the urgency of their work and its implications. A series of briefing papers presented at the start of the conference encouraged the scientists, in the words of leading US meteorologist William Kellogg, to "stick their necks out".

Some of the papers had been written by a rising generation of climate scientists keen to make their mark. Of 22 authors of papers eventually published in a book marking the conference, no fewer than four came from the Climatic Research Unit, including Phil Jones and its new director Tom Wigley.

By the end of the conference Thomas Malone, who had chaired the 1983 NRC report, had changed his mind about the need for more research before taking action. He had, he said afterwards, undergone "a reversal of an opinion I held a year or so ago. I believe it is timely to start on the sensitive task of framing a convention on greenhouse gases, climate change and energy."

Another factor in this change of perspective, says Franz, was that in Villach climate change stopped being the preserve of atmospheric physicists and modellers. Outsiders showed up, and they were shocked by what they heard. The engineers and biologists in attendance could see the immense practical significance of what was being forecast. It seemed to them, in the words of one, "very scary".

Some scientists who weren't there said later that Villach wasn't so important. Wigley, who despite authoring a paper hadn't attended, told me Villach was a "waffly non-event" whose significance has been "grossly overstated". It is true that, among climate scientists, the story of the significance of greenhouse gas emissions and their accumulation in the atmosphere had been building for some time. Stephen Schneider, one of the first climate modellers to take global warming seriously, reckons the important modelling work was done in the US in the 1970s.

But there was new science at Villach. Bolin presented analysis showing for the first time that it was not just carbon dioxide that was raising temperatures. Several other gases being pumped into the air were having a similar effect, including methane, nitrous oxide and chlorofluorocarbons (CFCs), the same gases that were destroying the ozone layer.

This revelation had a big impact on the participants, Bruce said later, because "it was bringing the likely time of significant changes in climate much closer than had been previously expected".

Most, however, agree that the change brought about by Villach was as much about personal as atmospheric chemistry. The meeting somehow generated the "wow factor" says Jaeger – the realisation that there could be significant change in our lifetimes. "While evidence for the importance of greenhouse gases had been mounting," she says, "it was this group of scientists who mobilised around the findings and used them to justify a call for urgent action."

This was the key moment when climate change went political. The bandwagon was rolling. Bruce and Tolba went home to use their senior positions at WMO and UNEP respectively to promote the case for establishing a scientific panel to assess climate change. The WMO formally adopted the proposal at its congress in 1987. Another participant, Gordon Goodman of the Beijer Institute in Stockholm, gave climate change a prominent place in a chapter on global energy issues that he was then writing for the influential Brundtland report on environment and development. That report led directly to the UN Earth Summit in Rio de Janeiro in 1992, where the UN Climate Change Convention, precursor to the Kyoto Protocol, was signed.

Meanwhile, other politically savvy scientists became attracted to the issue. Some, like the British atmospheric chemist Bob Watson, who went on to become the second chairman of the IPCC, had cut their policy-making teeth earlier in the decade persuading politicians to tackle the threat to the ozone layer.

It took another four years for global warming to fully grab the attention of the world. That happened one June day in 1988, in the midst of a major drought across the US. Jim Hansen of NASA's Goddard Institute for Space Studies testified to a hearing of the

Senate Energy and Natural Resources Committee in sweltering heat, and with the air conditioning mysteriously malfunctioning. He mopped his brow and declared to the senators that global warming "has reached a level such that we can ascribe with a high degree of confidence a cause-and-effect relationship between this greenhouse effect and observed warming". The story made front pages round the world: "Global warming is here and now."

A few days later, researchers meeting at a high-profile conference on the science of the atmosphere in Toronto demanded a 20 per cent cut in global greenhouse gas emissions by 2005 – a target that, much watered down, became the basis for the Kyoto Protocol of 1997. And before the year was out, Tolba and Bruce had persuaded the UN General Assembly to let them establish the Intergovernmental Panel on Climate Change to assess "the scientific, technical and socioeconomic information relevant for the understanding of the risk of human-induced climate change".

They didn't get it all their own way. Participants at the Villach meeting, particularly Tolba, had envisaged establishing an independent group of scientists to monitor the issue. But when officials from the US government's Department of Energy heard from their scientists what had been going on in Villach, they complained loudly about the conference's political recommendations. "The US government became alarmed that the climate agenda was being driven by a small group of non-governmental entities," says Jaeger. So the Reagan administration worked the UN corridors, lobbying instead for the creation of an intergovernmental body whose activities would be approved by government delegates.

The purpose of the IPCC, so far as Reagan's people were concerned, was to put scientists back in the cages they had escaped from at Villach. But it was too late. The story of global warming

– and what scientists really felt about it – was out. Far from knuckling under, the Villach generation of climate scientists and their heirs have used their new-found status as the experts on an intergovernmental panel to underline the urgency of tackling climate change, and sometimes strong-arm governments into signing off on their reports.

Republican US administrations since have devoted much time to denigrating the regular assessment reports of the IPCC. (Every five years or so, it produces volumes covering the science of climate change, its impacts and methods of mitigation.) But the irony is that it is a body that they did more than anyone to create. And the spirit of Villach lived on.

Some have seen this as a weakness in climate science ever since – a propensity to leap over the barriers of pure science into something akin to advocacy. And to massaging reports towards political ends. Others say that it is only the scientists who see the perils the world faces from climate change, and they are right to speak up.

CHAPTER 4

Playing hockey with climates past

"I know there is pressure to present a nice tidy story"
KEITH BRIFFA TO HIS IPCC BOSSES, SEPTEMBER 1999

It is a persuasive image. The "hockey stick" graph shows one estimate of the average global temperature over the past 1,000 years. For the first 900 years there is little variation, like the shaft of an ice-hockey stick. There is no sign of Hubert Lamb's medieval warm period or little ice age, just a flat shaft. Then, in the 20th century, comes a sharp rise like the stick's blade. The hockey stick has become one of the central elements in the public argument that what is happening to the world's climate today is unusual, maybe unique.

The IPCC put the graph in the summary of its third climate assessment, published in 2001. Although it was intended as an icon of global warming, the hockey stick has become something else – a symbol of the conflict between mainstream climate scientists and their critics. The sceptics have made it the focus of their attacks for a decade. Some hope that by demolishing the hockey stick graph they can destroy the credibility of climate scientists. And in the man who first drew the hockey stick, a young paleo-climatologist called Mike Mann, they have found an angry, outspoken and sometimes vulnerable foe.

In the battle over the hockey stick, both sides have retreated to hard-line positions. You either believe the hockey stick version of past temperatures, or you don't. You see it as climatic reality, or a cynical caricature drawn in the service of politics rather than truth. Quite illogically, believing in the hockey stick has become synonymous with believing in the threat posed by climate change. So mainstream scientists often feel they must defend it, whatever their private doubts. But doubts there were from the start.

The *Guardian* investigation of the emails discovered a vitriolic debate within the mainstream science community in 1999, during preparation of the IPCC report, about the validity of the graph. It is a debate that neither climate scientists nor the naysayers have discussed in public because it doesn't fit their simple two-tribe view of the climate wars. Mann and CRU's tree ring specialist Keith Briffa usually share common cause today, and they are portrayed by their enemies as co-conspirators, but the CRU emails reveal that back then they were in competing camps. Mann promoted his hockey stick; Briffa was very dubious, especially about the prominence the IPCC wanted to give it.

The stakes were high. In the late 1990s, the IPCC was preparing its third major assessment report of the state of climate science. The report's leading authors were keen to demonstrate the level of natural climatic variability. That would allow researchers to test whether recent warming had exceeded natural bounds. It could be proof that the "hand of man" was visible in current climate change. Thermometer records only went back a century and a half, however. So what was needed was some reliable proxy for temperatures further back in time.

In 1996, I visited Briffa at his lab at CRU for an article I was preparing on the use of tree rings as proxies for past temperatures. He told me: "Five years ago, the climate modellers wanted nothing to do with the paleo community [scientists studying past climate]. But now they realise they need our data. We can help them define natural variability."

For years, analysing the annual growth rings in ancient trees was an arcane discipline. Researchers such as Briffa knew that in hot summers, trees usually grew more, leaving wider and denser growth rings that could be dated by simply counting backwards from the bark. There were other influences on the growth rings, such as droughts. And no two trees are alike, let alone two forests. Nonetheless, all sorts of intriguing data had begun to emerge. Researchers found thin rings in trees around the world after major volcanic eruptions that caused a year or two of global cooling.

Back then Briffa was excited by a new discovery. The year 1032 was looking like the coldest on record in many places. Temperatures had plunged as a result of one of the largest volcanic eruptions in the past 10,000 years. Mount Changbai in China, close to today's border with Korea, blasted 50 cubic kilometres of rock into the air, darkening the skies and plunging the world into a wintry gloom. Carbon dating of trees buried by the eruption had put it at early in the 11th century, but Briffa's painstaking analysis of tree ring growth in the Urals had pinpointed the precise year of the quake.

But, as I reported in *New Scientist*, "it is not just climatic shocks that interest Briffa... Studying tree ring sequences across longer timescales can also reveal more general trends, including the natural pattern, range and variability of global temperature fluctuations. And that, Briffa believes, has the potential to answer one of the most vital questions of our time: has human activity really started to warm the planet?"

Despite Changbai, the 11th century overall looked rather warm in many places, partially fitting the ideas of Briffa's former boss

Hubert Lamb about the medieval warm period. But the picture was mixed. I wrote: "The 11th century seems to have been a particularly turbulent time in the Urals: 1045 was the warmest summer of the millennium, whilst 1032, the probable year of the Changbai eruption, was the coldest summer in 1,000 years. Within a single year however, temperatures had switched back, and 1033 produced the second warmest summer of the millennium at more than 2C above the mean." Further east in the Yamal peninsula, however, Briffa told me, he found no evidence of the medieval warming.

That contrast between the Urals and Yamal has come back to haunt Briffa. Some sceptics say he has more recently favoured the Yamal tree ring data over his Ural proxies because they fit the hockey stick shape. We will explore that issue in the next chapter. But when I interviewed him back in the 1990s, Briffa revelled in the diversity of tree ring chronologies. Such superficial agendas as hockey sticks were not for true tree ring aficionados.

At the same time that Briffa was developing his temperature proxies based on tree rings, other analysts were producing other kinds of proxy climate data. They were based on the size of glaciers and air bubbles trapped in ice, the isotope imprint left in coral reefs, the types of seeds in lake sediments and the temperature of water at different depths in deep boreholes.

Tim Barnett of the Scripps Institution of Oceanography, part of the University of California, San Diego, joined Phil Jones to form a small group within the IPCC to mine this data for signs of global warming. They planned to summarise the research in the panel's next assessment, due in 2001. They had an agenda. "What we hope is that the current patterns of temperature change prove distinctive, quite different from the patterns of natural variability in the past," Barnett told me in 1996. Even then they were looking for a hockey stick.

Up stepped Mann, then a young researcher at the University of Virginia – the same campus, incidentally, as the arch-climate sceptic Pat Michaels. He and colleagues Ray Bradley of the University of Massachusetts Amherst and Malcolm Hughes, a British-born tree ring specialist at the University of Arizona, began one of the first serious attempts to work out fluctuations in average global temperatures over the past millennium. Most tree ring records were from Europe and North America. So Mann's team tried to build a more global picture of the northern hemisphere by including proxies of different sorts from as many regions as possible.

It was pioneering work, assembling and collating data that had never been put together before. They published their first graph, showing average temperatures in the northern hemisphere going back to AD1400, in *Nature* in 1998. The following year the team extended the reconstruction back to AD1000. This 1999 version, published in the journal *Geophysical Research Letters*, was dubbed the "hockey stick" – not by Mann, but by Jerry Mahlmann of the National Centre for Atmospheric Research in Boulder, Colorado.

The long straight shaft of the hockey stick was a surprise. Conventional climate histories, including Lamb's famous study, pointed to a much more wavy line, with a warm period in the medieval period around AD1000, followed by a little ice age. Mann's explanation has always been that these phenomena were largely European and North Atlantic phenomena. They were not global. Indeed, he says, it was likely that if it was warmer in some places back then, it would be cooler in others.

He is not alone. Mike Hulme, formerly of CRU and a professor of climate change at the University of East Anglia for the past 20 years, is noted as an independent thinker on climate issues. He says: "Put simply and loudly, we don't know whether there was a global medieval warm period or not. It is not the case that prior to the hockey stick there was good scientific evidence for it. There just wasn't. What we had was Hubert Lamb's sketch on the back of an envelope, which was endlessly repeated. But it was based heavily on the central England temperature series and his reasoning/speculation. The hockey stick may be rubbish, too, but it was not overturning a well-established orthodoxy. It was replacing a rather dodgy bit of hand-waving from the 1970s and 1980s."

Even so, when Mann, Bradley and Hughes first produced their hockey stick, many tree ring researchers in particular doubted whether the graph had got it right. Initially, Mann and his co-authors expressed similar concerns. The title of their 1999 paper, "Northern hemisphere temperature during the past millennium: inferences, uncertainties and limitations", was hardly bombastic.

Reconstructing past temperatures from proxy data is fraught with danger. Tree ring records, the biggest component of the hockey stick record, sometimes reflect rain or drought rather than temperature. When I investigated the continuing row surrounding the hockey stick graph in 2006, Gordon Jacoby of Columbia University in New York said: "Mann has a [tree ring] series from central China that we believe is more a moisture signal than a temperature signal... He included it because he had a gap. That was a mistake and it made tree ring people angry." A large data set he used from bristlecone pines in the American west to fill another gap, pre-1400, has attracted similar concern.

Deciding which data sets to include in such reconstructions was, if not arbitrary, then open to dispute. And dispute there was. In the late 1990s, the researchers were in heated debate about what they could and could not reliably show about past temperatures, and how to represent their findings. The debate heated up as they came under pressure to deliver on Barnett's expectation of a "distinctive pattern" for the IPCC.

The hockey stick was certainly a distinctive pattern. The temptation to hype it was huge. As the stick began to appear in the scientific literature, Jones and his team of thermometer-watchers at CRU reported that 1998 was the warmest year in their 150-year record of data. The length of the hockey stick blade just grew. Those in charge of publicising the work of climate scientists and making the case for man-made climate change – often the same people – were understandably excited. Controversial science swiftly morphed into a propaganda tool.

The World Meteorological Organisation put the hockey stick on the cover of its 1999 report. Then IPCC chiefs decided to give it pride of place in their 2001 IPCC assessment. Based on Jones's data and the hockey stick, they stated in the summary for policy-makers that "it is likely that the 1990s have been the warmest decade and 1998 the warmest year during the past thousand years". That attracted attention – and trouble. The doubts expressed in the title of Mann's 1999 paper about "uncertainties and limitations" were becoming inconvenient as IPCC editors sought to present a simple, comprehensible message for a public audience. But was it good science?

Here the CRU emails reveal a previously untold story. Exchanges in September 1999 show intense disagreement about whether Mann's hockey stick should go into the IPCC summary for policy-makers – the only bit of the report that usually gets read outside the scientific community – or whether other reconstructions using tree ring data alone should get priority. The emails also expose the tensions between a desire for scrupulous honesty about uncertainties, and the desire for a simple story to tell the policy-makers. The IPCC's core job is to present a "consensus" on the science, but in this critical case there was no easy consensus.

The tensions were summed up in an email sent on 22 September 1999 by IPCC author Chris Folland, a researcher at Britain's Met Office Hadley Centre. He alerted key researchers that a diagram of temperature change over the past 1,000 years "is a clear favourite for the policy-makers' summary". But, he said, there were two competing graphs covering the full millennium and using a range of proxies – Mann's hockey stick and another, by Jones, Briffa and others. There was also a third candidate, a graph compiled by Briffa alone, using tree rings only and covering the period since 1400.

Mann's graph was clearly the more compelling image of man-made climate change. "The current diagram with the tree ring-only data somewhat contradicts the multiproxy curve and dilutes the message rather significantly," said Folland. "We want the truth. Mike [Mann] thinks it lies nearer his result (which seems in accord with what we know about worldwide mountain glaciers and, less clearly, suspect about solar variations)." Folland noted: "This is probably the most important issue to resolve in chapter two at present."

It looked like a clear thumbs-up for the hockey stick. But three hours after receiving Folland's message, Briffa sent a long and passionate email demanding caution over the use of Mann's hockey stick. "It should not be taken as read that Mike's series is THE CORRECT ONE," he warned. "I know there is pressure to present a nice tidy story as regards 'apparent unprecedented warming in a thousand years or more in the proxy data', but in reality the situation is not quite so simple... For the record, I believe that the recent warmth was probably matched about 1,000 years ago... and that there is strong evidence for major changes in climate over the Holocene [the period of the last 12,000 years] that require explanation and that could represent part of the current or future background variability of our climate." This last point is important. Briffa was saying not only that the hockey stick might not be right, but that any graph of the last 1,000 years could not be taken to represent the limits of natural variability.

The September spat was the latest manifestation of a simmering row. Only hints appear in the published emails. But they reveal intense anger behind the scenes. Earlier in the year, there had been a dust-up over an article by Briffa and his CRU colleague Tim Osborn submitted to *Science*. This discussed the uncertainties surrounding the hockey stick and similar reconstructions of past temperatures. Asked to comment prior to publication, Mann had sent an angry review of the article to the journal. We don't know precisely what he said, because the review was sent privately to Science and does not appear in the CRU emails, but he was clearly intemperate and appeared to have hoped to prevent its publication. Some of his friends thought that was over the top.

In April 1999, Ray Bradley, Mann's co-author on the hockey stick papers, broke ranks in an email. "I would like to disassociate myself from Mike Mann's view that 'xxxxxxxxxxx' and that they 'xxxxxxxxxxxxx'. I find this notion quite absurd. I have worked with the UEA group for 20+ years and have great respect for them. As for thinking that it is 'better that nothing appear, than something unacceptable to us'... as though we are the gatekeepers of all that is acceptable in the world of paleoclimatology [this] seems amazingly arrogant." (The apparent deletions are in the email as published.)

Days later, back from a holiday, Jones also laid into Mann: "You seem quite pissed off with us all in CRU. I am somewhat at a loss to understand why. It is clear from the emails that this relates to the emphasis placed on a few words/phrases in Keith/Tim's *Science* piece. These may not be fully resolved but the piece comes out tomorrow. I don't want to open more wounds but I might by the end of the email. I've not seen the censored email that Ray has mentioned, but this doesn't to my way of working seem the way you should be responding – ie slanging us all off to *Science*. We have disagreements, but… we have never resorted to slanging one another off to a journal (as in this case)."

The article by Briffa and Osborn, titled *Seeing the Wood for the Trees*, discussed in detail the reliability of different graphs of temperatures based on proxies. These were the core issues behind the IPCC's debates about the hockey stick. It is worth quoting at length from the article because many of the warnings it contains have been taken up since by critics of the hockey stick. Not many of these critics have pointed out that they started with Briffa.

The article warned that the reliance placed by Mann on "tree ring chronologies in the western United States" for interpreting temperature before 1400 may be misplaced. This was the bristlecone pine problem. "These trees display a progressive increase in growth from the middle of the 19th century, which may be wholly or partly due to rising atmospheric CO_2 levels," they say, rather than to temperature. "How can we distinguish the growth-promoting effects of warm temperatures from the possible influence of increasing CO_2 and perhaps even… nitrogenous pollution?" they asked, before dismissing Mann's effort to eliminate non-temperature influences.

The article went on to question whether the shape itself meant anything, pointing out that the upper bounds of the error bars on the hockey stick "are high enough to encompass even the upper limit of uncertainty associated with their 20th century temperature estimates (which run up to 1980)". Briffa and Osborn agreed with Mann that the 20th century had been "anomalously warm", but warned against concluding that the medieval era was not.

Finally, it warned against reading too much into the fact that other researchers, including Briffa and Jones, have produced not dissimilarly shaped graphs. "An uninformed reader would be forgiven for interpreting the similarity between the 1,000-year temperature curve of Mann *et al* and a variety of others... as strong corroboration of their general validity and, to some extent, this may well be so. Unfortunately, very few of the series are truly independent. There is a degree of common input to virtually every one, because there is still only a small number of long, well-dated, high-resolution proxy records."

All this amounted to a siren warning about the dangers of taking too much notice of the hockey stick at that time. Though dramatic, Briffa and Osborn were saying, the image proves nothing. It is easy to see why Mann reacted so angrily.

Mann and his US colleagues have recently characterised the row over the *Science* article as "completely separate" from the row over the use of the hockey stick in the IPCC report. Strictly this may be true. But they are connected because the Science article is about the reliability of the hockey stick, and hence how much reliance the IPCC should place on it. Jones also saw a wider professional rivalry being played out. He thought the row went back to the previous summer. "I would suspect that you've been unhappy about us coming out with a paper going back 1,000 years only a few months after your *Nature* paper." This was the proxy temperature reconstruction by Jones and Briffa.

Mann, Jones and Briffa eventually settled their differences. And the hockey stick was given pride of place in the IPCC report. Folland says: "My recollection is that the final version [of the IPCC summary], which contains the hockey stick, satisfied Keith and everyone else in the end – after the usual vigorous scientific debate." Up to a point. We will see Briffa's concerns about IPCC

hyping of the hockey stick return in debates about its 2007 assessment. But once the three came under attack from climate sceptics,
all reference to these past spats disappeared from the emails. They
faced a common foe. But many of the arguments used by Briffa
and Osborn in that 1999 article continue to underpin scepticism
about the reliability of the hockey stick.

CHAPTER 5
The Yamal tree ring "virus"

"Keith has been accused of fraud"

Trevor Davies on Keith Briffa, October 2009

In the weeks before the hack, one word dominated the CRU emails. That word was "Yamal". When I phoned Jones on the day the emails were published online and asked him what he thought was behind it, he said: "It's about Yamal, I think." The word turns up in 100 separate emails, more than "hockey stick" or any other totem of the climate wars. The emails began with it, back in 1996, and they ended with it.

The Yamal peninsula is one of the most remote places in the world, a frozen peninsula jutting into the Arctic Ocean from Siberia. The very first email in the leaked files, dated 6 March 1996, is from a Russian tree ring researcher called Stepan Shiyatov to Keith Briffa at CRU. Shiyatov, from the Institute of Plant and Animal Ecology in Yekaterinburg in the foothills of the Ural mountains, announced that he had got a small grant to take a helicopter to measure tree rings in timber hauled from frozen bogs in the permafrost of the Yamal peninsula. But they needed more money and hoped Briffa could help.

"It is important for us if you can transfer the advance money on the personal accounts which we gave you earlier," he wrote.

"Only in this case we can avoid big taxes and use money for our work as much as possible. Please, inform us what kind of documents and financial reports we must represent you and your administration for these money."

Briffa was red-hot for the idea. Large data sets of tree rings from outside Europe and North America were rare. Tree rings going back more than 1,000 years were gold dust. Two months later, Shiyatov was in Britain holding meetings on the project. The race to find the Yamal tree ring chronology, masterminded by Shiyatov and colleague Rashit Hantemirov, was born. That autumn, I was interviewing Briffa for a magazine article about his work. The draft of my article, which shows up in the emails, says: "On the Yamal peninsula, just east of the Ural mountains, wood dug from the permafrost grew in conditions so cold that some summers... no growth rings formed. Nevertheless, Yamal is the only site so far found that yields tree rings through a gap at 300BC."

Briffa and Shiyatov already had some results, published in *Nature* the previous year. They showed, I wrote, "unusual 20th-century summer warmth in a 1,000-year temperature record from Siberia. A complete tree ring chronology from AD914, pieced together from larches on the Yamal peninsula, suggested that average summer temperatures since 1901 have been higher than for any similar length of time during the chronology. It estimated that from 1600... to the present day there has been a 1.14C warming. The first eight decades of the 20th century were 0.13C warmer than the next warmest period, nine centuries before in 1202–91." The Yamal trees exhibit no medieval warm period of the kind seen in most European tree rings. It was thus an early pointer to the idea that medieval warming was far from being a global phenomenon, and that global temperatures may have followed a "hockey stick" trend.

In any event, Yamal was a treasure trove for paleoclimatologists. And Briffa pursued those tree rings tirelessly, but controversially. For contrarians, they have recently become a symbol of their disbelief and scepticism almost as great as the hockey stick. They have shouted foul from the rooftops. In the final weeks of the emails, correspondence over them grew to fever pitch, with the university pro vice-chancellor, the former CRU director Trevor Davies, saying the university should "pursue the [American] *Spectator* more vigorously. Keith has been accused of fraud."

It seems hard to believe that a bunch of tree trunks frozen deep in the Arctic more than two millennia ago could undermine the argument about man-made climate change. But in the politics of climate change anything is possible. In autumn 2009, Steve McIntyre was moved to suggest that one tree alone, known as YAD06, could be "the most influential tree in the world".

Briffa published data from Yamal regularly. It proved popular among analysts trying to piece together the history of climate change. Many of those studies showed up in the IPCC report published in 2007, where the relevant chapter was authored by Briffa. By then, McIntyre was on the trail of Briffa's research, chasing down the raw data and demanding to know why he chose to include certain sets of tree rings in his analyses and not others. Briffa said he had used data from trees that were long enough and "sensitive enough" to provide reliable data. In other words there were technical reasons. He denied cherry-picking data sets that exhibit a hockey stick shape. McIntyre has not specifically made that charge, though others have.

Even so, Briffa failed to release his raw Yamal data until, after one of his papers was published by the Royal Society in its *Philosophical Transactions* in 2008, editors required him to publish. Curiously, it only emerged in October 2009 that McIntyre had

had the Yamal data since 2004, when it was supplied to him by Russian researchers. He denies knowing this all along, however, stating on his blog: "I already had a version of the data from the Russians... What I didn't know until a couple of weeks ago was that this was the actual version that Briffa had used."

McIntyre's assiduous data-mining established that Briffa's analysis of the most recent warming, when the Yamal rings showed a strong hockey stick "blade", was based on just 12 trees. They became known as the "Yamal-12". McIntyre said this was far too small a sample, and other ring data was available to enlarge it. He found some and, hey presto, the hockey stick shape disappeared.

For Briffa, as a tree ring technician, what McIntyre was doing was farce. McIntyre simply didn't know what he was dealing with, he argued. The new data, Briffa said with evident disdain for the ignoramus on his trail, was estimating temperature from "maximum late wood density" measurements. These were quite different from the tree ring width measurements used in the Yamal series. Scientifically they were not comparable. And the data McIntyre had chosen only went back 100 years, which was useless for the longer time series Briffa was interested in. Briffa didn't say McIntyre's work was invalid, just different – and certainly not a refutation of his own.

This was turning into a sterile debate, fuelled by mistrust. McIntyre intimated that Briffa's choices of what data to use might be based not so much on technical considerations as on a desire to manipulate data to reproduce a hockey stick shape. Briffa assumed that McIntyre had no right to ask questions and that his agenda was entirely political. If anyone was cherry-picking data to fit his agenda, it was McIntyre. But while the two men locked horns in their emails, the demonology about Briffa was growing across the blogosphere and soon reached into mainstream media.

Ross Kaminsky, a columnist on the magazine *American Spectator*, claimed: "One implication, supported by Briffa's near-decade-long refusal to share his data, is that he cherry-picked the data set that supported the conclusion he wanted to find." McIntyre's sometime co-author Ross McKitrick wrote a vitriolic piece in his regular soapbox, the *Financial Post* in Canada, that attempted to use this local row about a few trees in the Arctic to undermine the entire hockey stick graph. He argued that the Yamal data was like a virus infecting numerous climate reconstructions. He charged that "the key ingredient in most of the studies that have been invoked to support the hockey stick, namely the Briffa Yamal series, depends on the influence of a woefully thin subsample of trees and the exclusion of readily available data for the same area".

What are we to make of this? Briffa's boss, Phil Jones says: "We would all accept that paleoclimatic data are considerably less certain than the instrumental data. However, we must use what data are available in order to look at the last 1,000 years. I believe that our current interpretation of the Yamal tree ring data in Siberia is sound. Yamal is just one series that enters some of the millennial-long reconstructions that are available."

Many scientists would agree with one British colleague of Briffa, who told me angrily: "Why should Briffa – one of the world leaders in this field – have to explain himself to people who are not even specialists in this area, who are in fact amateurs who have set themselves up as judges of professional activity and demand that scientists justify their actions to them?"

Briffa denies any wrongdoing and has attempted to explain his work. In September 2009, he put out a statement on the CRU website defending his research. It said: "We would never select or manipulate data in order to arrive at some preconceived or unrepresentative result." And there is nothing in the emails or anywhere else to suggest this isn't true. "We do not select tree core samples based on comparison with climate data. Chronologies are constructed independently and are subsequently compared with climate data to measure the association and quantify the reliability of using the tree ring data as a proxy for temperature variations."

The CRU evidence to the Muir Russell inquiry into Climategate says that any suggestion that Briffa and his colleagues "improperly selected certain tree ring data specifically in order to manufacture a false impression of recent enhanced growth in the Yamal region is… entirely false" and "defamatory".

A few days after Briffa posted his initial defence against McIntyre in September 2009, his old boss at CRU, Tom Wigley, was discussing the affair in an email to Jones: "Keith does seem to have got himself into a mess," Wigley said. He felt Briffa had not at that stage answered McIntyre's charges fully. "How does Keith explain the McIntyre plot that compares Yamal-12 with Yamal-all? And how does he explain the apparent 'selection' of the less well-replicated chronology rather than the later (better-replicated) chronology?"

Wigley offered some paternal criticism of the kind he had in the past offered to Jones, Santer and others on the inadvisability of withholding data. "Yes, there are reasons. But many good scientists appear to be unsympathetic to these. The trouble is that withholding data looks like hiding something, and hiding something means (in some eyes) that it is bogus science that is being hidden."

Three weeks later, Briffa and fellow CRU tree ring researcher Tom Melvin put up a longer justification addressing many of the points raised by Wigley. In February 2010, Wigley made clear that he was "completely satisfied" with this more detailed response.

"Not only did it answer all my concerns and questions, but it also shows that the real villain here is McIntyre," he said.

The Yamal data has become of great importance for scientists trying to analyse past climates. And Briffa is as vulnerable as any tree ring chronologist to the charge raised by Briffa himself against Mann's hockey stick a decade ago in his 1999 *Science* article, when he said: "Unfortunately, very few of the series are truly independent. There is a degree of common input to virtually every one, because there is still only a small number of long, well-dated, high-resolution proxy records."

It may be that, with more than a decade of effort invested into trying to extract data from the frozen wastes of Siberia, Briffa gave his Yamal rings too much prominence. But the truth is that any set of proxy data, whether tree rings or others, is only a small piece in a large jigsaw. Even if perfect, it shows a temperature record for just one place. Only by putting many of them together will a picture of wider trends emerge.

The danger is that if one data set gets used too much, it gains undue importance in creating the overall picture. Has this happened here? Maybe a bit. A handful of rings filling both a geographic and time gap in the record are immensely useful. But it is not true that the Yamal rings are an omnipresent virus in reconstructions of past temperature. McKitrick's op-ed claim that they are "the key ingredient in most of the studies that have been invoked to support the hockey stick" is poppycock.

For a start, they are not in the original hockey sticks at all. They were not published then. According to Jones in an email last October, of the dozen 1,000-year reconstructions of temperature examined during the last IPCC assessment, only three included Yamal data. And as the RealClimate website sarcastically noted, they are not in hockey stick graphs of temperature trends deduced from retreating glaciers, or water temperatures in boreholes, or core sunk into ice sheets. Take them away and the hockey stick still stands.

There are questions about how they have been used. These are questions that the small world of tree ring analysts would do well to address, as Wigley suggested. And the Oxburgh inquiry into the science of CRU, set up by the University of East Anglia in the wake of Climategate, took a similar view of Briffa's work. Reporting in mid-April 2010, it noted that "with very noisy data sets [such as tree rings] a great deal of judgment has to be used. Decisions have to be made on whether to omit pieces of data that appear to be aberrant. The potential for misleading results arising from selection bias is very great." This was inevitable, it agreed, but it created a strong "obligation to document the judgmental decisions". Sadly, "CRU accepts with hindsight that they should have devoted more attention in the past to archiving data and algorithms and recording exactly what they did."

Oddly, perhaps, Oxburgh accepted CRU's explanation that "at the time the work was done they had no idea that these data would assume the importance they have today". Some might question that. After all it was 14 years earlier that Briffa had told me in interview that "five years ago, the climate modellers wanted nothing to do with the paleo community [scientists studying past climate]. But now they realise they need our data. We can help them define natural variability." As early as 1996 he professed no doubt as to the new significance of his work in the wider story of climate change science.

Nonetheless the evidence is of sloppiness rather than anything more sinister. And the elevation of the Yamal tree rings into a totem of climate change "fraud" is nonsense. Even McIntyre appears sensitive to the charge that he feeds this process. Witness his response to a challenge on this point from RealClimate in the run-up to Climategate.

On 30 September 2009, RealClimate attacked those like McKitrick, the *American Spectator* magazine and *National Review* who suggest that Briffa is guilty of fraud over Yamal. Or, as *Daily Telegraph* blogger James Delingpole put it, "How the global warming industry is based on one MASSIVE lie". In his story the punchline revealed that the whole show – Al Gore, the hockey stick and all – was based on Briffa's supposedly dodgy tree rings from Yamal. RealClimate noted that McIntyre did not personally commit these rhetorical crimes. But it added sarcastically: "No doubt he has written to the *National Review* and the *Telegraph* to clarify their mistakes and we're confident that the corrections will appear any day now. Oh yes."

Well, McIntyre passed on the *Daily Telegraph* and the *National Review*. But, according to a blog he wrote a few days later, he had replied to the *American Spectator* saying: "While there is much to criticise in the handling of this [Yamal] data, the results do not in any way show that AGW [anthropogenic global warming] is a 'fraud', nor that this particular study was a 'fraud'. There are many serious scientists who are honestly concerned about AGW and your commentary... is unfair to them." Sadly, when last checked, there was no sign of this comment being published on the magazine website, though in January 2010 its editors had found room for another feature on the "great hoax" of climate change.

CHAPTER 6

Urban heat and a Chinese riddle

"This crowd of charlatans look for one little thing...
Best thing is to ignore them completely"
MIKE MANN ON PHIL JONES'S CRITICS, APRIL 2007

We all know that cities are hotter than the countryside, especially at night. The phenomenon was first identified in 1810 by the British amateur meteorologist Luke Howard. Cities generate heat. And the concrete, bricks and asphalt of urban areas absorb more heat than the fields and forests beyond the city limits. The result is "urban heat islands". For climate scientists this raises an important question. Are the rising mercury levels recorded at thousands of weather stations round the world often simply a result of this phenomenon – of thermometers that were once in the countryside gradually being engulfed by urban areas?

Phil Jones and five fellow researchers conducted a study back in 1990 to answer this question. Titled "Assessment of urbanisation effects in time series of surface air temperatures over land", their paper in *Nature* examined temperature data from three regions: European parts of the Soviet Union, eastern Australia and eastern China. Jones compared the temperature trends at rural and urban sites. He concluded that in each area, rural sites had warmed almost as much as urban ones. This confirmed his view that local urban heating was not a significant factor in the warming trend being recorded by thermometers round the world. It meant that the warming was truly global, that his graph of global temperatures measured by thermometers around the world over more than a century, the crown jewels of CRU's work, could be relied on.

Some of the most compelling evidence came from the study in eastern China, a densely populated agricultural region urbanising fast even then. If urbanisation influences were negligible there, they would likely be negligible everywhere. Since the 1990 study, most climate researchers have felt safe in ignoring the potential of urban heat islands to confound their temperature trends. The paper became a key reference source for the conclusions of succeeding reports of the IPCC. Most recently it featured strongly in a chapter in the 2007 assessment, a chapter co-authored by Jones. That chapter said that globally "the urbanisation influence… is, at most, an order of magnitude less than the warming seen on a century timescale". In other words, it is tiny. But how safe is that conclusion?

It is difficult to imagine a more bizarre academic dispute than that over the precise location of 42 weather monitoring stations in remote parts of rural China. But the argument over the weather stations of the Middle Kingdom, and how they affect an important set of data on global warming, has led to accusations of scientific fraud and may yet result in a significant revision of a scientific paper that is still cited by the UN's top climate science body.

The problem is that crucial data about the location of these Chinese weather stations used in Jones's paper cannot be verified, because documents containing it no longer exist. And the emails suggest that Jones helped cover up this embarrassing fact. As the world learned from Watergate and Richard Nixon, it is often the cover-up, rather than the crime itself, that gets you in the end.

Jones's Chinese data came from 84 weather stations, half rural and half urban, provided as a result of a joint project between the Chinese government and the US Department of Energy. In a country undergoing rapid urbanisation, it was critical to Jones's findings that the "rural" weather stations were genuinely remote from spreading urban areas. If they were too near to towns, and subject to creeping urbanisation, the distinction between rural and urban stations would be lost and the conclusion would be false. Location was all.

Climate sceptics have long argued that the data collected by Jones and others suggesting global warming from rising concentrations of greenhouse gases in the atmosphere could be polluted by this urbanisation effect. Lamb mentioned the problem in his 1984 British Association lecture. They have been convinced that the urban effect was much bigger over certain regions than that allowed for by Jones. It might not change the overall story of global warming too much, they have conceded. After all, two-thirds of the planet is covered by ocean, and the oceans are warming, too. But to ignore it is sloppy science.

In the case of the China study, the suspicion grew that Jones and his colleagues had not done their homework properly. Of particularly concern was the role of Wei-Chyung Wang, a Chinese-American climate researcher at the University at Albany in New York, who collated the Chinese data. Jones had attested that the Chinese stations had been chosen because they met high standards for the reliability of their data. Specifically they were "those with few, if any, changes in instrumentation, location or observation times". Yet McIntyre found that there were no published records of the precise locations of the 84 Chinese stations. And when he asked Jones to provide details about their locations, Jones refused him. He argued that it would be "unduly burdensome".

McIntyre says he spent two years asking Jones for the station location data, without success. Only following a freedom of information request in 2007 did Jones relent and put location data on the CRU website. He thought it would put an end to the affair. He couldn't have been more wrong. For the data – or rather its paucity – proved to McIntyre that his suspicions had been right all along: that Jones's data was a meteorological ragbag from which nothing could be proved either way about urbanisation influences on Chinese temperature trends.

The data supplied by Jones on the 84 weather stations in the study contained location information for only 35 stations. The other 49 had no histories of their location or other details. They were simply identification code numbers. These mysterious stations included 40 of the 42 rural stations whose remoteness was critical to the paper's findings. Of the remaining 35, the location data revealed that 18 had been moved during the study period, perhaps invalidating their data.

Enter Doug Keenan. The Canadian banker, now a naturalised Briton, had taken up amateur climate investigation and decided to dig into the affair. He asked a simple question: could it still be true, in the light of the data vacuum, that Jones and Wang had really chosen the stations to be included in their study on the basis of their impeccable station histories? Questioned, Wang said that the information certainly did exist at the time the stations were chosen – and he probably still had it somewhere.

But Keenan went back to the published documents from the joint US-Chinese programme that garnered the data in the first place. He told the *Guardian*: "Jones said they had got their data from a report that was published by the US Department of Energy and the Chinese Academy of Sciences. Yet a report from them published in 1991, just a year after Jones published his paper, said that for 49 of the 84 weather stations that Jones relied on, 'station histories are not currently available' and 'details regarding instrumentation, collection methods, changes in station location or observing times... are not known'." If the people from whom Jones and Wang got their data did not know this information, how could Jones and Wang? "Those statements imply that the essential claim by Jones *et al* is impossible." It amounted, he said, to fraud.

Keenan says the error was probably Wang's, but that Jones could not be exonerated. His literature search had turned up a paper that Jones wrote with Yan Zhongwei of the Institute of Atmospheric Physics in Beijing, in the journal *Advances in Atmospheric Sciences* in 2001. It described how two urban stations in China, one in Beijing and one in Shanghai, had been moved, and how this affected the temperatures they measured. The Beijing station had moved five times and over 41 kilometres. Its indentifying code showed it was one of the stations included in the 1990 study. So by 2001, Jones "must have known that the claims of Wang were not wholly true," he said. That charge assumed Jones had cross-checked the station codes. But Keenan demanded that Jones retract his claims about the selection of the Chinese stations and amend his paper's conclusions accordingly.

The emails reveal that when Keenan first made his allegations, Jones was hurt, angry and uncertain. "It is all malicious... I can live with the website abuse, but the Keenan letter knocked me back a bit. I seem to be a marked man now," he wrote in April 2007. Another email from him on the issue at that time said: "My problem is I don't know the best course of action. Just sitting tight at the moment taking soundings. I'd be far happier if they would write some papers and act in the normal way. I'd know how to respond to that. In a way this all seems a different form of attack from that on Ben [Santer] and Mike [Mann] in previous IPCCs...

I know I'm on the right side and honest, but I seem to be telling myself this more often recently!"

Several of his regular email correspondents gave him a shoulder to cry on, and suggested how he should respond. Mann advised him in forthright terms: "This crowd of charlatans... look for one little thing they can say is wrong, and thus generalise that the science is entirely compromised. The last thing you want to do is help them by feeding the fire. Best thing is to ignore them completely." Kevin Trenberth urged a fight back. "The response should try to somehow label these guys and [*sic*] lazy and incompetent and unable to do the huge amount of work it takes to construct such a database."

In August 2007, Keenan submitted a formal complaint about Wang to his employers at Albany. The university launched an investigation. Reporting in May 2008, it found "no evidence of the alleged fabrication of results and nothing that rises to the level of research misconduct having been committed by Dr Wang". The university's interim president, George M Philip, wrote Wang a fulsome letter declaring: "I concur there is no evidence whatsoever that you have committed data fabrication or any research misconduct. I have asked the vice-president to work with you to ensure that the university takes the appropriate steps to restore your scientific reputation."

But the university did not publish the detailed findings of the investigation, and refused to give a copy to Keenan. It justified this on the grounds that the report "did not include portions addressing your role and opinions". Since Keenan's complaint had been the genesis of the investigation, this was odd. It sounded like a kangaroo court in reverse.

Keenan had not been waiting for the inquiry outcome, however. Even as the university panel sat, he published his charges in the sceptics' house journal *Energy and Environment*, edited by Sonja Boehmer-Christiansen. His paper was entitled "Fraud allegations against some climate research of Wei-Chyung Wang". The allegations covered both Jones's 1990 paper and a parallel paper published in *Geophysical Research Letters* the same year, specifically on "Urban heat islands in China" by Wang and two colleagues. One was Zhaomei Zeng, a researcher at the Institute of Atmospheric Physics, part of the Chinese Academy of Sciences, and visiting scientist at the University at Albany in New York. The other was the head of NOAA's National Climate Data Centre, Tom Karl.

Keenan's paper was handled by Benny Peiser, a social anthropologist and lecturer at the School of Sport and Exercise Sciences in Liverpool's John Moores University. He took an interest in climate issues and was guest editor for a special issue on the activities of the IPCC. Peiser took the bull by the horns and asked Jones to comment on it. "If the claims are unsubstantiated, I would certainly reject the paper," he told him.

Jones swiftly emailed Wang and his old CRU boss Tom Wigley, saying he wouldn't be responding. But Wang said he should reply to deny the charge. "We are not going to let Keenan doing [*sic*] things his way... We should be thinking, after the odeal [*sic*] is over, to take legal (or other) actions." Jones replied that an action for libel in British courts might be a good idea. Mann got in on the act: "You should mention that the publisher of a libel is also liable for damages – that might make Sonja B-C be a little wary. Of course, if it does get published, maybe the resulting settlement would shut down E&E and Benny and Sonja all together! We can only hope, anyway."

A week later, Jones did respond to Peiser. He declared that the claims were "unsubstantiated" and said that the paper should not be published. But he did not repudiate the detailed claims. Keenan says another reviewer recommended publication, while a third reviewer brought in to adjudicate also approved the paper. At any event, Peiser notified Keenan and Jones that he intended to publish.

Keenan applauds Peiser and his small journal for championing the cause of an amateur. He told the House of Commons Climategate inquiry in 2010: "I doubt that mainstream science editors would have dared to reject the opposition by leading climate scientists who had targeted an amateur researcher." He justified going ahead with publication because Jones took no steps publicly to "clear up the discrepancies... It is unacceptable that the scientist who disseminates a data product on which international treaties are based... should actively seek to suppress information that calls the quality of the data into question."

The published paper was largely ignored at the time. This was probably because, if it was in any way untrue, it looked extremely libellous. But the emails reveal that, despite their angry bluster and legal threats, there was concern among some of Jones's colleagues about Wang's missing data on station locations – and the failure by Jones and Wang over several years to come clean.

In August 2007, when the Keenan draft paper was first circulated, Tom Wigley had warned Jones by email: "Seems to me that Keenan has a valid point. The statements in the papers that he quotes seem to be incorrect statements, and that someone (W-C W at the very least) must have known at the time that they were incorrect. Whether or not this makes a difference is not the issue here." Wigley said he took up the matter with Jones partly because he had been director of the CRU when the original paper was published in 1990. As he told Jones later, in May 2009: "I realise that Keenan is just a trouble-maker and out to waste time, so I apologise for continuing to waste your time on this, Phil. However, I *am* concerned because all this happened under my watch... and, although this is unlikely, the buck eventually should stop with me."

In a recent correspondence with Keenan, shown to the *Guardian* by Gavin Schmidt, Wigley says: "I can't remember the detail [but] it is likely that Phil convinced me that you didn't have a point at all." However, in his 2009 email, Wigley also put to Jones allegations made by Peiser in an email. "Wang had been claiming the existence of such exonerating documents [revealing the locations of the Chinese weather stations] for nearly a year, but he has not been able to produce them. Additionally, there was a report published in 1991 (with a second version in 1997) explicitly stating that no such documents exist. Moreover, the report was published as part of the Department of Energy Carbon Dioxide Research Program, and Wang was the Chief Scientist of that program." This was the DOE/CAS report referred to by Keenan, which obtained the original Chinese temperature data.

Wigley continued: "The next puzzle is why Wei-Chyung didn't make the hard copy information available. Either it does not exist, or he thought it was too much trouble to access and copy. My guess is that it does not exist – if it did then why was it not in the DOE report?"

Parts of Wang's defence to the university inquiry, seen by the *Guardian*, say that he had got the Chinese temperature data from his Chinese colleague, Zhaomei Zeng. Oddly, Zeng is not listed as one of the authors or collaborators on Jones's paper, though she was a co-author of Wang's paper on urban heat islands published in the same year. Wang revealed that in 2008, Zeng sent him a note explaining that she had lost her notes on many station locations during a series of office moves. Nonetheless "based on her recollections", she provided information on 41 of the 49 stations with no official written record.

Collating all the data, and assuming Zeng's recollections were correct, it seems that no fewer than 51 of the 84 stations in Jones's study moved during the 30-year study period, 25 did not move, and eight Zeng could not recollect. Wang, however, maintained to the university that the 1990 paper's claim that the stations were selected on the basis that they had "few, if any, changes in instrumentation, location or observation times" remained correct.

The inquiry had apparently accepted this version of events. Wigley was not so sure. In his May 2009 email to Jones, he said of Wang: "I have always thought W-C W was a rather sloppy scientist. I therefore would not be surprised if he screwed up here... Were you taking W-C W on trust?" This is a damaging charge against a senior researcher. Wang has not responded to it. Nonetheless, it is fair to say that there is no published evidence of any previous doubts being raised over Wang's work. Wang told the *Guardian*: "I have been exonerated by my university on all the charges. When we started on the paper we had all the station location details in order to identify our network, but we cannot find them any more. Some of the location changes were probably only a few metres, and where they were more we corrected for them." This is a somewhat circular argument since one purpose of the study would have been to work out how much correction might be needed for stations suffering urbanisation.

Wigley continued in his 2009 email to Jones: "Why, why, why did you and W-C W not simply say this right at the start? Perhaps it's not too late." Eight months later, Jones appeared to be coming round to this view. He continued to deny "hiding" flaws in the Chinese data. But when *Nature*, which had published the original paper two decades before, asked if he might submit a correction to the journal, he replied: "I will give that some thought. It's worthy of consideration."

He added a little ruefully: "The stations probably did move." Nonetheless, he told the Press Association, "I am confident in my own mind the site movements that might have taken place at some of the sites were not that important to affect the average of the 42 sites." *Nature* said he conceded the subsequent loss of data was sloppy. "It's not acceptable… not best practice."

All a fuss about nothing? Surely not. The purpose of the widely cited and hugely influential paper was to assess whether the precise location of weather stations mattered when assessing long-term temperature trends. So for Jones to say that he was confident in his own mind that it didn't matter negates the purpose of the study.

Meanwhile, there was a startling postscript. In 2008, Jones prepared a paper for the *Journal of Geophysical Research* on temperatures in eastern China. A re-examination of data from the original 84 stations used in the 1990 study affirmed that the 1990 analysis was correct. The results were "almost identical" says Jones. Given that the concern was over the reliability of the data rather than its analysis, that doesn't help a lot. But the paper also extended the study to look at a total of more than 700 stations in eastern China, and to update the data to cover the period since the early 1980s. And that new analysis found that the urban heat phenomenon was responsible for 40% of the warming seen in eastern China between 1951 and 2004. Far from being negligible, it made up almost half the warming.

This does not flatly contradict Jones's 1990 paper. The timeframe for the new analysis includes the period of fastest urbanisation in China since the 1980s. But it certainly reaffirms the original concerns of sceptics about the urban heat island effect. And it raises afresh their questions about one of the most widely referenced papers on urban influences on global warming, and about the IPCC's reliance over many years on its conclusions.

Jones, in his university's submission to the House of Commons Climategate inquiry, says of the 1990 study: "There was no attempt at misrepresentation." And the dramatic change in perception of the influence of urban heating in eastern China from his new paper "is simply an example of scientific research evolving as more and better data become available".

Gavin Schmidt at RealClimate argues that the row about the Chinese data is "absurd", and that Keenan's "knee-jerk presumption of misconduct when people disagree with you" has "no place in the scientific discourse, and serves only to poison scientific debate". To me, Keenan's charges of "fraud" and "fabrication" do seem extreme and unnecessary. But those of covering up flaws in research, certainly after 2001, are harder to refute. And the defence that it is all 20 years ago, and therefore irrelevant, is not tenable when the research concerned remains an underpinning of current reports of the IPCC.

It is important to keep the row, and the charges made against Jones, in perspective, however. Jones's revised estimate of the impact of urbanisation on temperatures in China is undoubtedly dramatic, and could presage similar reassessments in other fast-industrialising countries like India and Brazil. But it does not greatly change the global picture of temperature trends. There is plenty of evidence of global warming, not least from oceans far from urban influences. In a review of recent studies published online in December by the journal *Wiley Interdisciplinary Reviews: Climate Change*, David Parker of the Met Office concludes that, even allowing for Jones's new data, "global near-surface temperature trends have not been greatly affected by urban warming trends".

Keenan accepts that his allegations do not change the global picture. But he told the *Guardian*: "My interest in all this arises from concern about research integrity, rather than about global warming *per se*. Jones knew there were serious problems with the Chinese research, yet continued to rely upon the research in his work, including allowing it to be cited in the IPCC report."

PART TWO

The human element

CHAPTER 7

Scientists in the firing line

"There is a strategy to single out individuals, tarnish them and try to bring the whole of science into disrepute"
BEN SANTER ON THE IPCC'S CRITICS, 2006

One of the lessons of Climategate is that scientists are all too human. That discovery may be a relief to some, but it also gives an added dimension to battles fought out in the name of truth and enlightenment. The next few chapters concentrate on this human element. They expose how some climate scientists bent the rules of scientific engagement in ways that blurred the distinction between doing good science and preserving their own reputations. But we begin with an examination of some of their most determined critics and the very personal manner in which they sought to undermine the mainstream science.

Many of the researchers caught up in the Climategate saga have spent years in the firing line of sceptics. They bear the scars. The sceptics have a huge range of motivations, ranging from personality clashes and professional snobbery, through real differences of scientific opinion, to libertarian demands to "open up science". Some of the most vocal sceptics, however, are not interested in the science. They have more commercial and political agendas, furthering the aims of their employers in undermining calls for global action against greenhouse gas emissions. The scientific sceptics are few in number but the power often lies with the people from this political end of the spectrum – the fossil fuel companies and right-wing thinktanks and newspaper proprietors – who are keen to publicise their views. It is also the case that some physicists have contempt for climate scientists, such as a former president of the US National Academy of Sciences, Frederick Seitz, and Dick Lindzen, a distinguished meteorologist from the Massachusetts Institute of Technology, who told me in an interview in 1997 that the IPCC was "dominated by guys from the bottom of the heap, such as geographers".

Even so, among the IPCC scientists Lindzen has the best reputation of all the sceptics. British meteorologist John Houghton, who chaired three IPCC assessments of climate science, called Lindzen "the most serious scientist among the sceptics". In the emails, Tom Wigley noted his presence in 2005 as a "token sceptic" on a review group and added: "At least he is a smart guy and he does listen." Even Mike Mann called him "sharp", and meant it as a compliment.

Lindzen accepts that carbon dioxide will warm the planet, but believes that the initial warming will unleash negative feedbacks that will damp it down, whereas most climate modellers reckoned the feedbacks are positive, amplifying the initial warming three-fold or more.

Lindzen has had some disagreeable friends. I first came across him during the Earth Summit at Rio de Janeiro in 1992. He had been invited to address some of the city's more illustrious citizens, who were rallying in opposition to the environmentalists in town. His hosts were right-wing and militaristic. The country was only seven years removed from military dictatorship and it felt like it.

But Lindzen barely seemed to notice the chanting and regalia on show.

He later invited me to his suburban home outside Boston, where we discussed photography. He was an academic through and through. Lindzen flatly denied taking more than occasional lecture fees from corporations and fancy lobbyists, and I believed him. But he said that others were less fastidious. "Western Fuels does support two people heavily."

The Western Fuels Association represents a number of mostly coal-burning electricity utility companies in the American West. Pat Michaels of the University of Virginia received some of its largesse. When I met Michaels in Charlottesville in 1997, he immediately struck me as different from Lindzen. He was a sharp scientist but a much more political operator. After hearing an oral presentation from a PhD student in his lab, he headed off campus to an office where he ran his "advocacy science consulting firm", now trading as New Hope Environmental Services, which specialised in attacking certain mainstream climate scientists. Back then he had climate modeller Ben Santer of the Lawrence Livermore National Laboratory in his sights. He regularly visited Washington to talk climate politics with lobbyists at the George C Marshall Institute and elsewhere, and had persuaded a past governor of Virginia to give him the honorary title Virginia State Climatologist.

Like Lindzen, Michaels accepted that warming was happening, but he insisted that it would be a modest affair. And he expected the mainstreamers to come round. "The truth is that what we sceptics say is always pilloried by the climate modellers, and then adopted as their own five years later." At the time he was chafing that he had proposed in the early 1990s – in an article in *New Scientist* magazine – that particulate air pollution was masking warming from greenhouse gases, an old refrain from the 1970s.

He had initially been rebuffed. But by 1994 modellers at Britain's Met Office had adopted his ideas as a prime reason why warming was not as great as their models then predicted.

By 1997, Michaels's long-running newsletter *World Climate Report* was already in business, largely funded by the Western Fuels Association. Other publicly revealed industry funders for Michaels over the years include the German Coal Mining Association and the Edison Electric Institute, another body of electricity utilities. In 2005, it emerged that Michaels had also received hundreds of thousands of dollars in consultancy fees from the Intermountain Rural Electric Association, a coal-burning utility based in Colorado. A leaked letter from the company's general manager, Stanley Lewandowski, discussing the payments, said: "We believe it is necessary to support the scientific community that is willing to stand up against the alarmists."

Michaels was also open in accepting the ideological and financial embrace of right-wing Washington thinktanks. One of the earliest was the George C Marshall Institute, which during the Ronald Reagan presidency became a rallying point for star wars armaments, before moving into climate advocacy. It received funding from ExxonMobil and once paid a lobbying visit to London at the invitation of the British energy industry. Other free-marketeers later joined the climate sceptics' cause, notably the Cato Institute, where Michaels became a fellow, the Competitive Enterprise Institute (CEI) and the American Enterprise Institute.

These groups mostly devoted little attention to the science of climate change. When I investigated the CEI in 2005, it employed six specialists on climate change. Two were lawyers, one an economist, one a political scientist, one a graduate in business studies and one a mathematician. The team was headed by Myron Ebell, an energetic advocate for the climate policies of the

Bush administration. He was most famous in the UK at the time for a tirade on BBC radio in which he accused the government's then chief scientist David King of "knowing nothing about climate science". The CEI made television advertisements in 2006 arguing against climate change, one of which ended with the words, "Carbon dioxide: they call it pollution, we call it life." It too has received ExxonMobil funding.

What these groups appeared to care most about was opposing any state or UN-run efforts to halt greenhouse-gas emissions. Even in the 1990s they had cleverly combined the two in their denunciation of Al Gore, who had written passionately about environmental issues before being elected vice-president to Bill Clinton in 1992. Gore, they said, planned to use climate change as a vehicle for "global governance". Back in 1997, Michaels made the seemingly wild and unsubstantiated claim to me that Gore would use corrupt environmental journalists to peddle a fraudulent version of climate change and advance his campaign for the presidency.

The anti-Gore campaign played well. Even his hugely popular film, *An Inconvenient Truth*, which first awakened many Americans to climate change, did not put them off. Thirteen years later, Gore's face featured prominently on a poisonous "hide the decline" animation on YouTube attacking Mike Mann.

Among those in positions of real power who took up the case against Mann, Gore and the "climate conspiracy" was Republican James Inhofe, who for years has used his chairmanship of the Senate Environment and Public Works Committee to campaign against climate scientists. In a two-hour address on the floor of the Senate in July 2003, he argued: "With all of the hysteria, all of the fear, all of the phony science, could it be that man-made global warming is the greatest hoax ever perpetrated on the American people? It sure sounds like it."

But the lobbyists and senators had a hard time finding many scientists willing to back up their rhetoric. Most of those they had were getting rather old. Like Sherwood Idso, a plant scientist from Arizona convinced that extra carbon dioxide in the atmosphere could only be good for the world; and the already legendary Fred Singer, who was convinced that only the sun seriously influences global climate and anticipated a new ice age any time.

The lobbyists wanted fresh faces to sell their case. And, most of all, they needed credible scientists. In 1998, I revealed in the *Guardian* leaked documents showing that the powerful American Petroleum Institute was planning to recruit a team of "independent scientists" to do battle against climatologists on global warming. The aim was to bolster a campaign to prevent the US legislature ratifying the Kyoto Protocol, which the administration had signed the previous year after a visit to the Kyoto conference by Al Gore.

The Institute's eight-page Global Climate Science Communications Plan said it aimed to change the US political climate so that "those promoting the Kyoto treaty on the basis of extant science appear to be out of touch with reality". The leaked document continued: "If we can show that science does not support the Kyoto treaty… this puts the US in a stronger moral position and frees its negotiators from the need to make concessions as a defence against perceived selfish economic concerns."

The plan's first task was to "identify, recruit and train a team of five independent scientists to participate in media outreach. These will be individuals who do not have a long history of visibility and/or participation in the climate change debate." It sounded like something out of a John le Carré spy novel.

It is not clear if the plan went ahead, or whether these moles were ever put in place in the scientific community. But the existence of such a plot has understandably led to a huge amount of conjecture among climate scientists about whether any of their more sceptical colleagues have been co-opted to this cause. Many believed the moles have been at work for years and that climate scientists are victims of a commercially funded conspiracy to destroy their work and scientific reputations.

In late 2006, I interviewed a number of scientists scarred by their battles with the sceptics for an article on how the propaganda war was shaping up prior to the publication of the next IPCC assessment the following year.

Santer had been attacked by Michaels and others for his role in writing the 1995 IPCC report, which claimed to see the hand of man in climate change. He said: "There is a strategy to single out individuals, tarnish them and try to bring the whole of science into disrepute." Mike Mann of Penn State University, fresh from battles over his hockey stick's appearance in the 2001 IPCC report, said: "There is an orchestrated campaign against the IPCC." Kevin Trenberth had suffered abuse for publicly linking global warming to the exceptional 2005 Atlantic hurricane season, which culminated in Hurricane Katrina. He told me: "The attacks on me are clearly designed to get me fired or to resign."

But there was also evidence that the scientists had been making trouble for themselves, by sometimes stumbling across the line between policy-relevant research and advocacy. Arguably Trenberth was one of those.

When Hurricane Katrina hit New Orleans in August 2005, it was not just the city that was rocked. Within weeks, climate science was embroiled too as a group of scientists, led by Trenberth, argued that the growing number of strong hurricanes in the Atlantic and perhaps elsewhere was down to global warming. After all, hurricanes only form when sea surface temperatures exceed 26C, and global warming meant that was happening ever more often.

Trenberth said recent changes in hurricane incidence were "much more likely to be related to global warming" than natural fluctuations. He was not alone. "We can say with confidence that the trends in sea-surface temperatures and hurricane intensity are connected to climate change," said Judy Curry of the Georgia Institute of Technology. She was co-author of a paper on how the number of hurricanes had doubled in 35 years – probably due to rising sea temperatures. It was, by chance, published within days of Katrina hitting.

Over at the RealClimate website, set up as a mouthpiece for climate scientists, Mann and others penned a statement asking, "Could New Orleans be the first major US city ravaged by human-caused climate change?" It conceded that you couldn't argue that for sure, but climate change had "loaded the dice" in favour of such disasters. This was instantly controversial, especially in a country being run by an administration that had reneged on the Kyoto Protocol. *Nature* reported that the White House had blocked the release of a statement earlier that summer from a panel of its climate scientists making the link between climate change and worsening hurricanes.

But other scientists argued that the links were not so clear-cut. One prominent hurricane forecaster, William Gray of Colorado State University, insisted that climate change made no appreciable difference to hurricanes and any data to the contrary was not to be believed. As a meteorologist caught up in short-term phenomena, he might not have spotted, or felt the use for studying, longer-term climatic trends. But there were also valid scientific reasons for thinking that warmer waters would not necessarily lead to more hurricanes. Some argued that what mattered was the difference between ocean temperatures and those high in the atmosphere, at the top of potential storms. So if the atmosphere was warming too, then nothing much would change.

That debate, which continues, was science as it should be conducted. But personalities kept getting in the way. Trenberth had been the first senior climate scientist to make the link between climate change and hurricanes. He had done so the year before Katrina, during a press conference, where he said historical records proved the link. At the time, Trenberth was working on the next report of the IPCC, and his claim infuriated an IPCC colleague, government meteorologist Chris Landsea of NOAA's hurricane research laboratory in Florida.

Landsea believed that measurements made before the era of satellite observation were not reliable enough to make the claim stick. He regarded Trenberth's claim as evidence of scientific bias not befitting someone then engaged in an IPCC assessment of the science of the matter. When Trenberth persisted, and denied he was doing more than lay out the science, Landsea resigned from the IPCC, saying it had been "subverted, its neutrality lost".

Trenberth was a marked man among climate sceptics. Weeks before Katrina hit, in June 2005, he told *Nature* in an interview: "I am trying to get people to think about things in a different fashion... The point is that all meteorological events around the world are influenced in some way by global warming." Pat Michaels concluded in a blog posted on his *World Climate Report* website the following day: "Obviously Trenberth has an agenda, and it is one that lies outside of the current state of our best science." Another sceptic, Roger Pielke Jr of the University of Colorado in Boulder, accused the IPCC leadership of seeing their role as political advocates rather than honest brokers. After Katrina, these science world spats for a while became headline news.

By 2008, the row had calmed down. So had the hurricanes. A new study from Thomas Knutson of NASA's Geophysical Fluid Dynamics Laboratory said warming of the Atlantic would likely result in fewer hurricanes in the Atlantic, though the worst ones would probably be stronger – a decidedly mixed blessing. As Curry told *Discover* magazine in early 2010, "Six months later, we had sorted things out and were talking to scientists on the other side of the debate. We ended up making pretty good progress on the hurricane story as a result."

"Compare that," she added, "with the hockey stick story, where there's been a war for six years running."

CHAPTER 8
Breaking the hockey stick

"The goddam guy is a slick talker and super-confident.
He won't listen to anyone else"
Wallace Broecker on Mike Mann, 2006

Some climate scientists responded to sometimes politically inspired antagonism to their work by taking the fight to the enemy. Among the most pugilistic was Mike Mann, purveyor of the hockey stick. His self-confidence and abrasiveness ensured that, for good or ill, the battles over the hockey stick would become a central feature of the climate wars, as played out in both the world's media and the privacy of the emails.

After the publication of the IPCC's third assessment report in 2001, the controversy about the hockey stick spread beyond the science community. Political opponents of climate scientists cried foul. What had been essentially an argument between Mike Mann and Keith Briffa, conducted within the IPCC system, turned into something much more toxic, with towering egos and the politics of UN action on climate change all brought into play. Mann, no shrinking violet, was in the thick of it. It has defined his career in ways he could never have expected.

Republican senator James Inhofe of Oklahoma, who in July 2003 told the Senate that global warming was a "hoax", has repeatedly attacked the Penn State professor by name, accusing Mann of "flawed research" and "bad science". "In obliterating the medieval warming period and the little ice age, Mann's hockey stick just doesn't pass muster," he told senators in January 2005.

In 2005, Congressman Joe Barton of Texas, chairman of the House Committee on Energy and Commerce, joined in. He ordered Mann, Bradley and Hughes to provide his committee with extensive details of their working procedures, computer programs and past funding. *Nature* noted acerbically that Barton's grasp of the issues was slight. He "has chosen to cherry-pick selected information on the hockey stick work, using an article from the *Wall Street Journal* as his scientific guide".

But Inhofe and Barton were less interested in science than in politics, the researchers believed. Tom Crowley, a paleoclimatologist then at Duke University in North Carolina, voiced a common view when he wrote in *Eos*, the newsletter of the American Geophysical Union: "At some point one must ask why should a member of US Congress get involved in this matter... The presumed logic is that if the Mann *et al* reconstruction can be proved flawed, then so too is the general report of the IPCC... guilty by association."

Mann has always been a tasty target. His voluble style has made enemies and exasperated his friends. "The goddam guy is a slick talker and super-confident. He won't listen to anyone else," one of climate science's most senior figures, Wally Broecker of the Lamont-Doherty Earth Observatory at Columbia University in New York, told me. "I don't trust people like that. A lot of the data sets he uses are shitty, you know. They are just not up to what he is trying to do... If anyone deserves to get hit, it is goddam Mann." (It should be said that Broecker too is a controversial character. He has a reputation among some scientists for bad-mouthing young researchers, though the man himself insists: "If you check around, you will find that I am extremely supportive of young researchers. The reality is that I have absolutely no patience with bad science, whether it be by junior or senior people.")

Briffa took a deep breath when I asked him in 2005 about Mann. "Mike," he said, "is honest but can get a bit overenthusiastic about his own work." That enthusiasm can get out of hand. Some accuse him of conducting a campaign of belligerence against his critics that has spectacularly backfired in the Climategate scandal. "The hockey stick thing was exacerbated by Michael Mann's behaviour," Judy Curry of the Georgia Institute of Technology told *Discover* magazine. "We've just seen this blow up and blow up and blow up." Mann responded to Curry's criticism with customary vigour that maybe ended up making Curry's point for her. "Did you ask Judith to turn over her emails for the past three years? Until she does that she is not in a position to be talking about other scientists. Glass houses. Look, I'll just say this. I've received emails from Judith that she would not want to be made public."

The temperature of the hockey stick debate soared in 2003 with the intervention of mathematician Steve McIntyre and economist Ross McKitrick. This initiative began with some posts on a website for climate sceptics in which McIntyre described what he said were attempts to recreate Mann's 1998 hockey stick, using Mann's statistical methods of analysing proxy data.

McIntyre knew about stats, but his co-author knew the ropes when it came to academic publishing. McKitrick helped McIntyre turn this analysis into a paper fit for publication. He also brought to the collaboration contacts in the fossil fuel industry and other organisations fanning the flames of climate scepticism, such as the

Fraser Institute in Canada and the Competitive Enterprise Institute in Washington DC.

Most of the claims about McIntyre's supposed links to fossil fuel interests date to this time and to his collaboration with the more overtly political McKitrick. On trips to Washington, the pair met Senator Inhofe and toured the thinktanks. And they chose as a likely outlet for their paper the avowedly political *Energy and Environment* journal. Its editor since 1998 has been Sonja Boehmer-Christiansen, a geography professor at the University of Hull.

I first came across her when she was questioning established concerns about acid rain back in the 1980s. She has always regarded environmental controversies as essentially political. She told the House of Commons inquiry into Climategate in 2010 that "the IPCC consensus is politically created in order to support energy technology and scientific agendas that in essence pre-existed the 'warming as man-made catastrophe alarm' " and that CRU scientists "acted in good faith, doing their duty to science, bureaucracy and the public as they saw it and as they were funded to do". But she felt entitled to respond, through her journal, "to give sceptics a voice". Boehmer-Christiansen said later she rushed the paper by McIntyre and McKitrick into print "for policy impact reasons, eg publication well before COP9". COP9 was the latest round of climate negotiations, held in Milan that December.

In *Energy and Environment*, McIntyre for the first time publicly called into question the statistical methods employed by Mann to amalgamate his different data sets, and suggested that the hockey stick was an almost inevitable consequence of those methods.

This was not the kind of thing that Mann would take lying down. The leaked emails reveal that he heard about the paper that would cause him so much trouble only the day before it was published in October 2003. He was angry that the journal had not asked him to review the paper, or at least comment on it, before publication. Certainly that would have been normal etiquette. He put his friends on attack alert, sending a round-robin email: "My suggested response is to dismiss this as a stunt appearing in a 'journal' already known to have defied standard practices of peer review. It is clear, for example, that nobody we know has been asked to 'review' this so-called paper." He went on: "The claim is nonsense… Who knows what sleight of hand the authors have pulled? Of course the usual suspects are going to try to peddle this crap. The important thing is to deny that this has any intellectual credibility whatsoever."

In an ironic twist, Mann appended to this email the anonymous note that had alerted him to the paper, apparently after being copied among several scientists. The note said that, far from being nonsense, the paper reveals what "was known by most people who understand Mann's methodology [that] it can be quite sensitive to the input data in the early centuries". It went on: "There's going to be a lot of noise about this one, and knowing Mann's very thin skin, I am afraid he will react strongly, unless he has learned (as I hope he has) from the past…"

No chance.

While McKitrick went back to his teaching in 2004, McIntyre continued his analysis. The website he then ran, climate2003.com, discussed his efforts to get a critique of the hockey stick into a bigger journal, *Nature*. McIntyre, it seems, was starting to have fun. But I fear that journalists like me, who should by now have been taking this story seriously, were fooled by a mixture of the complexity of the statistical argument being made by McIntyre, and by the loud assertions by Mann and others that there was nothing in it.

So what was his case? There were a series of overlapping issues, but I will focus on three. The first criticism applies to all temperature reconstructions involving tree rings. While openly discussed in the scientific literature, it tends not to make it into public presentations. Tree ring analysts are pretty sure that from the mid-19th century, when they first have useable thermometer data, through to the mid-20th century, the width of rings faithfully represents real temperatures. Some detail is lost, but the overall measure is good. But since around 1960, tree rings do not grow in line with the warming measured by thermometers.

The cause of this "divergence" remains a troubling mystery. Some say it is probably a result of some other human-caused factor, like pollution with nitrogen. Others that it is some artifact of the process of measurement and analysis of tree rings. CRU's evidence to the Muir Russell inquiry said that the problem only arises in "a particular set of chronologies" made using tree ring data compiled by one team of Swiss researchers led by Fritz Schweingruber, a leader in the field. The evidence pointed out that none of these data sets extended back to the medieval era. It looks unlikely that whatever is causing the divergence has corrupted the proxy data 1,000 years ago, but until the cause is established nobody can be sure. (It would, by the way, be perverse to suggest that tree ring specialists at CRU or elsewhere have attempted to hide this problem. Keith Briffa, in a paper in 1998, was the first researcher to bring attention to it in the scientific literature.)

Second, McIntyre complains that the method that Mann pioneered to bring different data sets together in one graph – in which different proxy records are merged, sorted and their trends smoothed – had the effect of artificially flattening the hockey stick shaft. Any graph of real temperatures would have been much less smooth. That would have been a reasonable approach if all the data had been subjected to the same smoothing. But Mann had decided to add a trace of real temperatures onto the end, to cover the final decades, when tree rings were known to be unreliable. It gave a misleading impression, because there was no smoothing in this real data. The shaft had been smoothed, but the blade had not.

Mann didn't try to hide the technique he used in his papers. He put in error bars above and below the main line on his graph, showing how much temperature change the smoothing might have removed. He was proud to be among the first paleoclimatologists to put error bars on his reconstructions. The technique is one reason the error bars were so wide. As Briffa and Osborn pointed out back in 1999, most of the "blade" of 20th century warming would have fitted within the errors. It wasn't Mann's fault that in future renditions, those very wide error bars sometimes disappeared.

A third criticism, the main subject of McIntyre's proposed paper for *Nature*, was that Mann's analysis of temperatures contained a crucial mathematical flaw. To produce the graph, he measured how much data diverged from the 20th century mean, when he should have looked for divergence from the long-term mean throughout the 1,000 years. McIntyre pointed out that this was not just a technical error. It skewed the whole study, because it meant that almost any set of data would tend to generate a hockey stick shape. Mann didn't like it, but he eventually conceded the point in a correction published in the journal in July 2004. Humiliatingly after all the invective, it began: "It has been drawn to our attention (S McIntyre and R McKitrick) that the listing of the 'proxy' data set…"

McIntyre's paper never made it into *Nature*. But he and McKitrick finally got their extended analysis published in another journal, *Geophysical Research Letters*, in February 2005. This time,

McKitrick's friends stoked up the PR campaign in earnest. An essentially obscure paper in a second-ranked journal was splashed across the front page of Canada's *National Post* with a political spin: "A pivotal global warming study central to the Kyoto Protocol contains serious flaws." In news terms, the timing could not have been better. The protocol was set to come into force later that month. And it allowed the Bush administration, which had reneged on the protocol after coming into office, to claim the deal had been discredited.

The story grew. The *Wall Street Journal* put McIntyre on the front page, while Mann was accused elsewhere of fraud. For most climate scientists, confronted by headlines in papers that rarely reported their findings, it looked like a set-up. But the emails reveal that Mann was skating on very thin ice. There was growing private unease even among climate scientists who would defend him publicly. In October 2004, Tom Wigley had written a note to Jones, probably about the upcoming GRL paper. "I have just read the M&M stuff criticising MBH [Mann, Bradley and Hughes]. A lot of it seems valid to me. At the very least MBH is a very sloppy piece of work – an opinion I have held for some time. Presumably what you have done with Keith [Briffa] is better? – or is it? I get asked about this a lot. Could you give me a brief heads up? Mike [Mann] is too deep into this to be helpful."

Certainly Mann's emails were vitriolic in the extreme about M&M. In one sent to Jones, he described McIntyre as "this fraud... I would NOT RESPOND to this guy. As you know, only bad things can come of that. The last thing this guy cares about is honest debate – he is funded by the same people as [Fred] Singer, Michaels etc."

Despite Wigley's warnings, Jones sided with Mann unreservedly. He replied to Wigley that the McIntyre paper was "a complete distortion of the facts. M&M are completely wrong in virtually everything they say or do." The "bottom line", he added, was that there was "no way the medieval warm period was as warm globally as the last 20 years". Then, extraordinarily, he added: "This is all gut feeling, no science, but years of experience of dealing with global scales and variability."

In the aftermath of the attacks on Mann, a number of researchers have scrutinised the competing claims more coolly – and with more science. Hans von Storch, director of the Institute for Coastal Research, in Geesthacht, Germany, concluded that McIntyre and McKitrick were right to say that temperatures should be analysed relative to the 1,000-year mean, not the 20th-century mean. But he also found that even when this was done, it did not have much effect on the result.

Meanwhile, two people closer to Mann – his former student Caspar Ammann of the National Centre for Atmospheric Research in Boulder, Colorado, and Eugene Wahl of Alfred University, New York – claimed to have reproduced the findings of Mann and his colleagues for the period since AD1400. They concluded in Stephen Schneider's journal *Climate Change* that most of the difference between Mann and McIntyre over the shape of the hockey stick concerned the choice of which proxy data was included. They said that McIntyre had not really "repeated" Mann's study at all, as he had claimed. In fact he had done a different study, leaving out some of the sets of tree ring data that Mann included.

In particular, McIntyre had excluded the problematic bristle-cone pines in the south-west of the US. "Basically, the M&M case boiled down to whether selected North American tree rings should have been included, and not that there was a mathe-matical flaw in Mann's analysis," Ammann told me in 2006.

Interestingly, McKitrick now appears to agree. In a newspaper article in the Canadian *Financial Post* in October 2009, while still complaining that Mann's statistical methods skewed the data, he said of the hockey stick: "Its shape was determined by suspect bristlecone tree ring data."

Back in 2005, the paper by Wahl and Ammann was assuming considerable importance even before it was published. The emails show the IPCC authors knew all about it and were keen for it to get into print in time to be cited in the upcoming IPCC report. Nobody, it seems, wanted to backtrack on the claims made so prominently in the previous report about the unique warmth of the final years of the 20th century. But in the face of the McIntyre critique, they badly needed Ammann and Wahl's peer-reviewed confirmation of Mann's findings.

Jones was one of the reviewers who apparently gave the Wahl and Ammann paper a clean bill of health in May 2005. "Should keep them quiet for a while," he noted in an email to Mann that month, referring to M&M. But other reviewers seem to have been less happy and Schneider called for a revision. The deadline for acceptance of papers into the IPCC assessment process slipped from May 2005 to December and finally February 2006, when the paper was finally accepted for publication. The paper, "Robustness of the Mann, Bradley, Hughes reconstruction of Northern Hemisphere surface temperatures: Examination of criticisms based on the nature and processing of proxy climate evidence", did not finally appear in the journal until November 2007 – almost a year after the IPCC report itself was launched.

Keith Briffa was one of the lead authors of the IPCC chapter on paleoclimate. However, he was not the source of the drive to make full use of the paper by Wahl and Ammann and to downplay McIntyre's concerns. The emails show rather that, much as in

1999, he was keen not to hype the hockey stick. In a series of messages to his coordinating lead author, Jonathan Overpeck of the University of Arizona, in February 2006, he said: "We should be careful not to push the conclusions beyond what we can securely justify... Let us not over-egg the pudding... Do not let Susan [Solomon] or Mike [Mann] push you (us) beyond where we know is right." He was also corresponding with Wahl about the IPCC text in a manner that McKitrick, in his submission to the Muir Russell inquiry, claimed contravened the IPCC's rules for reviewing draft text. These emails were among those that, as we shall see later, his boss Phil Jones appeared to want to get deleted. But whatever the procedural correctness of what Briffa was doing, the evidence is that his aim was to get the report right in its nuances. Behind the complex statistical issues Briffa more than anyone was aware of the old problems with tree rings. In particular, the troublingly arbitrary nature of temperature reconstructions when the choices made about which data to include and which not seem often to be based on researchers' hunches, or the simple necessity to fill gaps in the chronology.

In 2006, as drafting of the IPCC report reached its final stages, the US National Academy of Sciences published the results of a long inquiry into the hockey stick, triggered by a request from Congress. It upheld most of Mann's findings, albeit with some important caveats. It agreed that there were statistical failings of the kind highlighted by M&M, but again it found that they had little effect on the overall result.

On the big question of the stick itself, the NAS study concluded: "There is sufficient evidence... of past surface temperatures to say with a high level of confidence that the last few decades of the 20th century were warmer than any comparable period in the last 400 years. Less confidence can be placed in proxy-based reconstructions of surface temperatures for AD900 to 1600, although the available proxy evidence does indicate that many locations were warmer during the past 25 years than during any other 25-year period since 900." Briffa would have agreed.

One panel member, Kurt Cuffey, a geographer from the University of California at Berkeley, reserved his criticism for the way the graph had been used by the IPCC back in 2001. "I think that sent a very misleading message about how resolved this part of the scientific research was," he said. And the Oxburgh report on CRU's science, published in April 2010, agreed. Papers published by Briffa and others at CRU emphasised the drawbacks of tree ring proxies and underlined that they could never be as good as thermometers, it said, "but presentations of the work by the IPCC and others have sometimes neglected to highlight this issue", something it regarded as "regrettable".

In retrospect, Mann kind of agrees. "Given its place in the IPCC summary with the uncertainties not even shown, we were a target from the beginning," he admitted to me.

Mann has always accepted that his graph was work in progress, and most researchers in the field accept that he is honest if hot-headed. "I'm not slamming what he did overall. It was a great effort, a great step," Gordon Jacoby of Columbia University told me in 2005. "But he got into hot water by defending it too hard in places where he shouldn't."

What counts in science, however, is not a single study. It is whether findings can be replicated by others. Here Mann has been on a winning streak. Upwards of a dozen studies, using different statistical techniques or different combinations of proxy records, have produced reconstructions broadly similar to the original hockey stick. These reconstructions all, crudely, have a hockey stick shaft and blade. While the shaft is not always as flat as Mann's version, it is present. Almost all support the main claim in the IPCC summary: that the 1990s was probably the warmest decade globally for 1,000 years.

McIntyre and McKitrick are unimpressed by this consensus. McIntyre says: "There is a distinct possibility that researchers have either purposefully or subconsciously selected series with the hockey stick shape." McKitrick similarly insists that there is a cabal of paleoclimatologists who have their favourite data sets that produce the required shape. In the *Financial Post* he singled out the US bristlecone pines and the Yamal tree rings. He said they occurred in so many studies that they skewed the lot. In fact, as we have seen, the Yamal tree rings were not in the famous hockey sticks of the late 1990s. Most of their data was not even published then. According to Jones, of the 12 reconstructions of temperature over the past 1,000 years used in the last IPCC assessment, authored by Briffa and published in 2007, only three contained Yamal data.

Whatever the reliability of the tree rings, other proxies that do not rely on trees show much the same. For instance, studies of the water in hundreds of wells and deep boreholes in all six continents, using formulae for how fast changes in temperatures at the surface conduct downwards, revealed that on each continent temperatures warmed faster during the 20th century than in any of the previous four centuries.

What are we to make of all this? Mann a decade ago was a young man in a hurry, a fiery character who made enemies and could sometimes intimidate his friends. He hasn't changed a lot. His ability to dominate a swathe of scientific debate almost by the force of his personality is troubling. He has sometimes lacked friends or colleagues able to rein in either his impetuosity or his temper.

Conflicts that could and should have been resolved instead festered and, in some cases, burst with the publication of the emails.

The hockey stick as drawn by Mann has taken some big hits. This pioneering piece of "work in progress" became victim of the notoriety it gained from being included in the IPCC summary in 2001. And of course to its catchy title. "The label was always a caricature and it became a stick to beat us with," Mann said later. Was it flawed research? Yes. Was it hyped by the IPCC? Yes. But while the McIntyre analysis has often been hailed as a refutation of the hockey stick, other analyses of proxy data using other statistical methods, including those without tree rings, have tended to confirm it. As Folland says today: "The Mann work still stands."

It also stands tall, because there is certain genius in the hockey stick. If the job of scientists is to make sense of confusing reality with simple statements and images that we can all understand, then Mann and his hockey stick may be remembered long after his critics are forgotten.

CHAPTER 9

Discerning the hand of man

*"In my more than 60 years... I have never witnessed a
more disturbing corruption of the peer review process"*

FREDERICK SEITZ IN THE *WALL STREET JOURNAL*, JUNE 1996

One man who has battled against climate sceptics longer than
most is Ben Santer. The American completed his PhD in climate
science in 1983 at CRU, where his boss was Tom Wigley and his
fellow academics included Phil Jones and Keith Briffa. Then he
headed for the Lawrence Livermore National Laboratory in California to become a career climate modeller. From there, he helped
write the second IPCC assessment report, published in 1995.
Most famously, this report found that "the balance of evidence
suggests a discernible human influence on global climate". It was
the headline-grabbing new finding of that report, and a break with
the caution of the panel's first report five years before. The story
of what happened next plumbs new depths in the character assassination of a scientist.

The IPCC reached its conclusion because, it said, the pattern
of warming across the globe closely matched that predicted by
models. For instance, warming at the Earth's surface was most
intense at high latitudes, and was matched by the expected cooling

in the stratosphere. Santer was the lead author of some of the key research behind this claim. He also wrote the relevant IPCC chapter. This left him unusually exposed to charges of manipulating the process – charges he has always vehemently denied. In fact, the record shows Santer spent a good deal of his authorial time ensuring that the chapter's findings were not over-interpreted.

The main body of the IPCC report, as written by Santer, included lots of qualifications about the claim to see the "discernible human influence". Another lead author on the critical chapter, Tim Barnett of the Scripps Institution of Oceanography, part of the University of California, San Diego, told me in 1996 that he and Santer "wrote a long list of caveats. We got a lot of static from within IPCC, from people who wanted to water down and delete some of those caveats. We had to work very hard to keep them all in."

While most of the caveats did stay put in the main body of the report, some did not make it to the summary for policy-makers, the bit of the report that outsiders are most likely to read. The *New York Times* leaked the final text with the headline: "Scientists finally confirm human role in global warming." It was a great headline. But the scientists must have expected trouble. They certainly got it.

Santer's co-authors on the critical research paper justifying the conclusions included several of his old CRU colleagues, including Jones and Wigley. But as the main author, Santer found himself under instant attack. Remembering all this a decade later, he told me: "I was accused of politically motivated tampering with the IPCC assessment, and of irregularities in my own research. I had, they said, somehow forced *Nature* to publish [my research] under duress, and had falsified scientific documents. There were calls for my dismissal."

His attackers were heavy hitters with often political agendas. Foremost among them were physicist Frederick Seitz, a president of the National Academy of Sciences in the 1960s, and Pat Michaels of the University of Virginia. Michaels and Santer go back a long way. Their battles have continued for a decade and a half. When I checked Michaels's *World Climate Report*, which is now largely web-based, one day while writing this, the top item was a comment about Santer's response on RealClimate to my articles on Climategate in the *Guardian*. Back in 1997, when I first met Michaels for an interview on campus in Charlottesville, he was spitting blood about... Ben Santer.

The charge from Seitz and Michaels back then was that Santer and the modellers kept fine-tuning their models so they fitted the available stats. For some researchers, particularly modellers themselves, this is a legitimate method of improving the models to get at the truth of how the climate system works. For Seitz and Michaels it was tantamount to fraud. Particularly if the modellers then came out of their labs to tell the world that their models must be right because they faithfully represent how the world is – which is what the IPCC report had just done.

"They can't go on tinkering with their models trying to make them fit reality," Michaels told me. He argued that the constant tweaking was aimed at covering up the fact that the climate was not warming as fast as the models predicted.

Michaels went further. Not only was Santer tinkering with his model, he was manipulating the data too. Or rather, being too clever by half in deciding which data to use in his analyses and which to ignore. He covered his desk in a series of graphs to demonstrate for me how, in the critical paper for the IPCC, Santer had ended his analysis of real-world global warming patterns in 1987. This was a significant year. It was the year when a long surge in warming in the southern hemisphere, relative to the northern hemisphere, went into reverse.

Michaels claimed that while the southern hemisphere warming fitted nicely with Santer's model predictions, the reversal in 1987 did not. He argued that if Santer had included data from the next couple of years in his comparison between model predictions and reality, it would have undermined the conclusion that there was a "discernible human influence" on temperature trends.

I put this to Santer later. He told me that "Michaels had a legitimate scientific concern about the sensitivity of our results to the choice of data period". But he emphatically denied any "sinister purpose" in his decisions to stop the analysis in 1987. That was simply the last date for which he had the data processed in the manner that he needed. After the IPCC report was published, and with a row brewing over its conclusion, Santer said he extended the analysis using later data. The findings were published in *Nature* at the end of 1996. Rather than undermining the case, it "strengthened the evidence for a human fingerprint", he said.

The science behind these differing interpretations is complex. But essentially it depends on how you evaluate different competing influences on hemispheric warming and cooling. On the one hand, oceans warm more slowly than the land, and the southern hemisphere is dominated by oceans. So you would expect it to warm more slowly. On the other hand, the shading effect of short-lived pollution aerosols like sulphates from human activity comes mainly from the north and has little effect in the southern hemisphere, tending to keep the north cool and let the south warm faster.

Santer said his model showed a clean-up of air pollution in the north resulting in a switch in the late 1980s. Michaels shrugged his shoulders and said the modellers had been tinkering with their models once again to make sure they came up with the "right" answer. But it is fair to say that more than a decade on, as Michael McCracken of the Climate Institute in Washington DC, notes, "Santer's analysis has held up; that of Michaels has not."

Amid the scientific arguments about who was right, Michaels's charges gained political traction. Santer, and by implications his co-authors Jones and Wigley, were labelled as frauds and cheats. Writing in the *Wall Street Journal*, Seitz, who later chaired the George C Marshall Institute, said: "In my more than 60 years as a member of the American scientific community, I have never witnessed a more disturbing corruption of the peer review process than the events that led to this IPCC report."

Arthur B Robinson, a biochemist from Oregon, claimed that Santer and his co-authors had "deliberately omitted data points to create the trend that they reported... So Santer clearly faked the result, circulated it during IPCC proceedings in order to influence world global climate policy. They should never be permitted to work in science again."

Robinson is an oddball. He is a Christian fundamentalist and has written a book about how to survive a nuclear war. He is also founder, president and professor of chemistry at a tiny outfit known as the Oregon Institute of Science and Medicine. With Seitz, he later masterminded the Oregon Petition, which declared against prevailing views on climate change and was signed by more than 30,000 people describing themselves as scientists. The two of them were out to discredit Santer.

Wearing his other hat as IPCC author, Santer also found himself accused of being the man who personally added the key words "discernible human influence" to the body of the IPCC report, and of doing it very late in the day. This was messy business and does not always reflect particularly well on the IPCC. But Santer looks like the fall guy here.

Those words were agreed at a main "plenary" session of the IPCC held in Madrid in November 1995, attended by government delegates from 96 countries as well as scientific authors. This final sign-off process with government delegates in attendance is why the IPCC has the word "intergovernmental" in its title. But Santer says the scientists "were there to ensure that the politics did not get ahead of the science".

It was a hard meeting, as civil servants and scientists grappled to find common ground. Discussions on Santer's chapter "were often difficult and contentious", he said. "We wrestled with the exact wording of the 'balance of evidence' statement. The delegations from Saudi Arabia and Kuwait argued for a very weak statement or for no statement at all. Delegates from many other countries countered that there was strong scientific evidence of a pronounced human effect." Industry lobbyists from the Global Climate Coalition, representing the interests of the American oil and automobile industries, were also heavily involved on the fringes of the meeting, briefing those who wanted to water down the findings. Santer says these lobbyists "dominated" initial discussions.

That plenary was one of the most intense and important debates in the IPCC's history. "After hours of often rancorous debate," Santer says, the then chairman of the IPCC, the affable Swedish meteorologist Bert Bolin, "suggested that the human effect on climate should be described as 'discernible'." The idea was accepted with open arms by exhausted delegates in the early hours of the morning and long after the UN translators had gone home. It was adopted in the summary for policy-makers. McCracken, who was a US government delegate in Madrid, says that the final outcome "reflects well on the IPCC".

But that wasn't the end of the matter for Santer. In fact, as an author of the report's main chapter on this issue, it made things harder. IPCC procedure requires that the chapters have to be made consistent with the summary, rather than vice versa. This is because the ultimate authors of the "intergovernmental" reports are the governments. But they only get to read and approve the summary for policy-makers. So if the summary says something different from the chapter it supposedly summarises, then it is the chapter that has to be changed.

Clearly such a rule potentially puts the scientists, as chapter authors, in a difficult position. And after the plenary meeting was over, Santer had the unenviable job of rewording his original chapter to reflect the political summary. He was also in charge of ensuring that all his co-authors were in agreement. So, as Santer put it in a discussion of what happened published on RealClimate, "Did the tail wag the dog?"

The charge made by Santer's critics is that the phrase about a "discernible human influence" went beyond what was said in the chapter from which the summary was supposedly drawn. And that, whatever had been agreed at the plenary, Santer was wrong to change the chapter to reflect the wording in the summary. Santer adamantly denies both charges. He says the summary was "an accurate reflection of the science. Had it been otherwise, I would not have agreed to put my name on the report." McCracken agrees.

In a way both sides are right and both sides are wrong. The chapter was a disaster waiting to happen, because the version brought to the plenary was far from being a finished, or even consistent, document. Santer points out in his analysis of the affair on RealClimate, posted in the wake of the *Guardian*'s reporting of this affair in February 2010, that the crucial chapter had been dogged by problems. He was only appointed as lead author for the chapter very late, after several people had turned it down. When he and his fellow authors arrived at the plenary, the chapter text was at a much earlier stage than other chapters. Santer says he flew into Madrid with a sheaf of review comments awaiting his attention. The existing draft was evidently a dog's breakfast. It contained sentences put in by different people at different stages that appeared contradictory. And, unlike other chapters, it also contrived to have both an executive summary and a "concluding summary".

Santer agrees that, in the end, he added the summary's words about a "discernible human influence" to his chapter late on, and without full consultation. But he said they amounted to "essentially the same conclusion" as that reached by the authors of the chapter months earlier. "The pre-Madrid bottom-line statement was: 'Taken together, these results point towards a human influence on climate.' Is 'suggests' stronger than 'points towards'? I doubt it. Is 'the balance of evidence' a more confident phrase than 'taken together'? I don't think so."

So far, so reasonable. But others point to other passages that disappeared from the report, either after Madrid or before. One passage asked: "When will the detection and unambiguous attribution of human-induced climate change occur? We do not know." Other statements stressing the uncertainties also hit the cutting room floor.

The contradictions are more apparent than real, however. For instance, it is much harder to show an "unambiguous" human impact than to pass a "balance of evidence" test. So it is not inconsistent to say that the balance of evidence points to a discernible human influence while at the same time agreeing that there is no unambiguous evidence of the same thing. But of course the emphasis is different. One can generate a headline like the *New York Times*'s "Scientists finally confirm human role in global warming", while the other will not. McCracken says sceptics "had planned some sort of public relations campaigns based around the failure to find 'unambiguous' detection of the human fingerprint".

The battle was about presentation as much as anything. And it went to the heart of the IPCC process. It was also extremely unpleasant. For Santer, the most disturbing language came from the lobbyists from the industry-funded Global Climate Coalition, who were aggrieved at not getting their own way during the Madrid plenary. In a report published in 1996, the coalition described Santer's final edit of the IPCC chapter as "scientific cleansing". The phrase was widely quoted. At the time it was a very obvious, and charged, reference to the ethnic cleansing going on in the Balkans. And for Santer, a Jew, it had another connotation. He told me in an interview I published in 2000: "My grandparents were subjected to ethnic cleansing. They died in a concentration camp in the second world war."

Santer spent months attempting to defend his reputation. He later discussed the barrage of invective he faced in terms similar to those used by Phil Jones after Climategate: "Nothing in your training prepares you for it. We are prepared for explaining our science, defending our science, and having scientists try to take your arguments apart. But we are not prepared for having our motives questioned and being accused of falsifying data. I think it is unproductive to engage with them [critics] directly. For many of them it is religious in a way. They are not rational. Don't waste our time; they don't have the same value system." The experience coloured Santer's world ever since. Colleagues say it contributed to the break-up of his marriage.

The Madrid plenary and the first battles over its outcome occurred a few months before the first of the leaked CRU emails was written. But later correspondence reveals Santer sharing those experiences with other victims of hectoring and abuse by the more rabid climate sceptics. Others had their own horror stories, including Mike Mann over his hockey stick graph, Kevin Trenberth over his analysis of hurricanes and warming in the aftermath of Katrina, and later Jones over his escalating data wars. In each case, they argue, their opponents turned sometimes legitimate debates about scientific analysis and data into vindictive character assassination. A recurring theme of the CRU emails is how the researchers sought to avoid falling victims again.

CHAPTER 10
Satellites, stats and bogeyman Santer

"They are best regarded as plausible cranks
who do not know what they are doing"

GRAHAM COGLEY ON DAVID DOUGLASS AND JOHN CHRISTY, MARCH 2010

In November 2008, Ben Santer believed he was being dragged back into the front line. He received a freedom of information request from Steve McIntyre asking for data connecting with a newly published paper. At least one senior colleague and co-author on the paper in question thought Santer would be best advised to hand over the data. But Santer decided instead to fight the request.

Santer's new paper was another major climate modelling study of the kind that he had made his life's work, cross-checking model predictions with the real climate. Titled "Consistency of modelled and observed temperature trends in the tropical troposphere", it had been published in the *International Journal of Climatology* the previous month. It reconciled an apparent contradiction between surface temperatures as measured by Jones's thermometer network and estimates of temperatures in the lower atmosphere, or troposphere, made by a series of satellites since 1979.

The contradiction had been debated for many years. While surface thermometers showed consistent warming, satellite data suggested the warming did not extend up into the atmosphere. This was unexpected, since climate models predicted the warming should penetrate upwards, especially in the tropics. It threatened to undermine Santer's case for finding a "discernible human influence" on temperatures.

Recent studies, however, had found problems with the analysis of the satellite data. Insufficient allowance had been made for the decaying of their orbit with time. The decay resulted in the software analysing the data from the satellite computing average temperatures in the air beneath that were too low. This suggested that the surface data was sound after all. But the next step was to see if the amended findings for temperatures aloft were consistent with the models. The priority for doing this had been increased by a report for the US government published in April 2006, which had highlighted the contradictory data as a "potentially serious inconsistency" in the science of climate change.

There were 17 authors involved in Santer's 2008 paper, including his old collaborators Jones and Wigley. Many of them hoped to lay the troubling issue of the satellite temperature estimates to rest once and for all. But while they were assembling their paper, Santer received from the journal *Geophysical Research Letters* a copy of a rival paper, written by David Douglass, a veteran astrophysicist at the University of Rochester, New York, with John Christy of the University of Alabama in Huntsville and others.

Douglass had, in a long career, looked at the energy balance on the Earth and written papers on chaos theory before taking up climate change. Christy had spent 10 years when young as a Baptist minister in Kenya before taking up science. By the 1980s he was cataloguing data from measurements of temperature obtained by US satellites, and was among the first to analyse why they were not showing either what the surface thermometers were recording or what the models were predicting. His unhelpful observations swiftly got him labelled a sceptic by some climate scientists. Meanwhile, he was adopted by the sceptics. When I first met him in 1997, he was equally discomforted by both. He reacted angrily when I suggested in an article that he was outside the mainstream. "There is nobody more mainstream in this than me," he wrote.

Christy and Douglass had collaborated on papers before, and clashed with Santer before. In a series of exchanges, Santer had sought to show that his models were consistent with the satellite data. Christy, often in collaboration with Douglass, had suggested this was wishful thinking. They had slugged it out regularly. In 2003, they had contrived to publish competing papers in the same week. In 2006 Douglass and Christy circulated a new paper for publication. Editors gave Santer a copy for peer review. Santer said it contained "serious scientific flaws". He recommended that the journal concerned reject it. Which it did, in September 2006.

Douglass persisted and produced a new version of his findings, which was published online by the *International Journal of Climatology* just over a year later in December 2007, ahead of an anticipated publication in the print version of the journal. This was increasingly standard practice among journals. The online paper was widely publicised by media that rarely report new science but are keen to question mainstream climate science. Fox News reported it, for instance. Douglass spoke about the paper at the National Press Club in Washington. Happy to feed a headline, he said it presented "an inconvenient truth" about climate change, which proved that "nature rules the climate. Human-produced greenhouse gases are not responsible for global warming." The right-wing Heartland Institute took up the argument.

The emailers, ever alert to their work being criticised in this way, discussed the affair. Jones agreed with Santer's judgment that the paper was statistically flawed. In an April 2007 email he wrote, "I know editors have difficulty finding reviewers, but letting this one pass is awful."

Jones and Santer had reasons for their concern about Douglass's statistical methods. The argument came down to a matter of statistical theory. The core of the matter was whether Douglass had properly used a statistical test to decide whether two sets of numbers – observed temperatures and modelled temperatures – were significantly different. The test was critical. If it found that the observed temperatures and those predicted by Santer's models were different, then that mean that Santer's model was wrong. And that in turn raised questions about whether he was right to see a human fingerprint in climate change. If the observed and modelled temperatures were sufficiently similar, as judged by the test, then that made Santer's model "consistent" with the real world – and gave everyone greater confidence in his wider claim about the human fingerprint.

Douglass said his application of the statistical test showed that the observed and Santer's modelled temperatures were significantly different. But Santer and Jones said he had misapplied the test, and if he had done it correctly there would have been no significant difference.

Who was right? This gets technical, but according to independent statisticians spoken to by the *Guardian*, the bottom line is that Santer was right and Douglass was wrong. Graham Cogley, a geographer at Ontario Trent University, explained: "There comes a point during the test where you have to divide by the square root of a number called the 'effective sample size'." In general, this number is smaller than the sample size because the samples are not independent of each other and this has to be allowed for. Cogley says that in his paper, Santer used the 'effective sample size' correctly. This produced wide error bars round the two sets of data, allowing him to conclude that there was no significant difference between the modelled temperatures and actual temperatures.

But Douglass fell into what Cogley calls a "trap". He divided instead by the square root of the sample size. He did not allow for the fact that the data in the sample were not fully independent. The result was narrow error bars. As a result, the error bars around the modelled and actual temperatures did not intersect. Cogley says the mistake that Douglass made is common. In this case it was critical, because it resulted in a false conclusion. (Cogley, it should be noted, is no patsy for mainstream IPCC scientists. He was the scientist who brought to the world's attention the false claim in the last IPCC report that the Himalayan glaciers would be all gone by 2035.)

Santer was angry that Douglass would not accept his mistake. The two sides went to war. But here the dispute becomes one of publishing protocol rather than science. And while it appears that Santer was right about the science, his hard-ball playing of the publishing system invites more criticism.

It was too late for Santer to do anything about the online publication of Douglass's paper by the *International Journal of Climatology*, or the publicity it had received. But Santer did not want it to go unchallenged when it finally appeared in the print version. So he buttonholed Jones's long-standing colleague at CRU, Tim Osborn. Osborn was a member of the editorial board of the journal. In an email to Osborn in January 2008, Santer said that Douglass and his co-authors were guilty of "intellectual dishonesty". Besides the statistical errors, he said that "Douglass *et al* ignored radiosonde data showing substantial warming of the tropical troposphere – data that were in accord with model results – even though such data were in their possession. Such behaviour constitutes intellectual dishonesty. I strongly believe that leaving these guys the last word is inherently unfair."

Santer told Osborn that a "quick publication of a response... would go some way to setting the record straight. I am troubled, however, by the very real possibility that Douglass *et al* will have the last word." To avoid that, he suggested, "our paper should be regarded as an independent contribution." Santer decided to redo much of Douglass's analysis using the correct statistical techniques.

Osborn in turn contacted the journal editor, Glenn McGregor, a climatologist at the University of Auckland. Osborn later told Santer that McGregor "may be able to hold back the [print version] of Douglass *et al*, possibly so that any accepted Santer *et al* comment could appear alongside it... Glenn suggested that I could help (because he is in Kathmandu at present) with achieving the quick turn-around time by identifying in advance reviewers who are both suitable and available."

In a follow-up note the next day, Osborn told Santer: "The only thing I didn't want to make more generally known was the suggestion that print publication of Douglass *et al* might be delayed." This might sound like skulduggery. But Osborn defended what he did in written evidence to the House of Commons Climategate inquiry. He said any suggestion that he used his editorial position "to give undue favour to Santer" was false.

In any event, from his distant vantage point in Kathmandu, McGregor was probably somewhat nonplussed by the flow of emails. One of the people copied into the emails from those organising the response to Douglass was someone who had reviewed the Douglass paper for him – and had failed to raise any objection. Nonetheless, McGregor agreed to a plan in which Santer would produce a response to Douglass as a formal paper – and the print version of the Douglass paper would be held back to run at the same time.

McGregor was as good as his word. Santer's paper was published online, with 16 co-authors, in October 2008. And the two papers appeared together in the same print edition of the journal the following month. So, though both papers took about four months from submission to publication online, Douglass's paper took 11 months to get from online to print publication, while Santer's paper managed it in 36 days. McGregor may well have regarded this as a correct response to a likely error in the Douglass paper, or simply seen it as a service to his readers to publish both competing papers together. But nobody told Douglass and his colleagues about any of this behind-the-scenes activity.

A year later, when the emails between Santer, Osborn and McGregor found their way into the public domain, Douglass and Christy reacted angrily to what seemed to them subterfuge. They sought the columns of the *American Thinker*, a daily internet magazine that lists its main subject matter as "national security in all its dimensions [and] the survival of the State of Israel". The magazine published the views of the pair on what they saw as a surreptitious strategy involving "delaying [our paper] and not allowing [us] to have a simultaneous response to Santer *et al.*"

In many ways this is a classic, if dispiriting, tale in the continuing battles over climate science. All the available evidence is that Santer was right about the science. But his influence over the process of scientific publication does raise questions. Santer denies the charge of conspiring to delay publication of the Douglass paper. But the emails provide strong evidence that he and his colleagues sought to use the influence they had to the utmost, albeit in defence of good science. On the other hand, whatever Santer's attempts to stage-manage publication, they were as nothing compared with the stage-management of Douglass's paper in the media. It gained far more, and far more prominent, coverage than Santer's paper.

In the world of science, Santer's team had the last word. Their charge that the statistical analysis in Douglass's paper was badly flawed and led to incorrect conclusions has, so far as the *Guardian* can establish, not been refuted, or even challenged. Cogley said of Douglass and Christy in early 2010: "They have had 15 months in which to respond to the criticisms of Santer and co, and all they have produced so far is whingeing about the editorial process. Until we see a reasoned, pertinent and convincing response from Douglass *et al*, they are best regarded as plausible cranks who do not know what they are doing."

But Douglass got the publicity.

That is where the story stands now. But it may not be the end of this continuing, festering argument. For, with the Santer paper finally published at the end of 2008, another controversial figure entered the fray – none other than Steve McIntyre. In late 2008, the Canadian mathematician asked Santer for data from the 49 computer model runs that he and his co-authors had conducted while preparing their paper for the *International Journal of Climatology*. He wanted to redo the analysis and see if he came up with the same answer. McIntyre says Santer turned down his request; Santer denies withholding anything. In his narrative of the affair on RealClimate, Santer says he told McIntyre that all the data he needed to replicate the analysis was freely available at his laboratory's website.

Things were, of course, a bit more complicated, as an email from Santer to McIntyre on 10 November 2008 reveals. "I gather that your intent is to 'audit' the findings of our recently published paper in the *International Journal of Climatology*. You are of course free to do so… You should have no problem in accessing exactly the same model and observational data sets that we employed. You will need to do a little work in order to calculate synthetic

Microwave Sounding Unit (MSU) temperatures from climate model atmospheric temperature information. This should not pose any difficulties for you… You will also need to calculate spatially averaged temperature changes from the gridded model and observational data. Again, that should not be too taxing. In summary, you have access to all the raw information that you require. I see no reason why I should do your work for you, and provide you with derived quantities which you can easily compute yourself."

This probably sounded like hard work – though it was what Douglass had done, albeit apparently making a mess of the statistical test of the findings. Santer had provided McIntyre with the information he needed to replicate his study, but he wasn't going to make it easy. McIntyre responded with a formal request to Tom Karl at the National Climate Data Centre, where he guessed some of the intermediate calculations made by Santer and his team would have been done. And he sent another request for the calculations to the journal, saying Santer's response had been "discourteous".

McIntyre also put in a freedom of information application for emails relating to the work between the 17 co-authors. He wrote on his blog 10 days after getting Santer's response: "If Santer wants to try this kind of stunt, I've submitted FOI requests and we'll see what they turn up. We'll see what the journal policies require…" McIntyre was testing the system. Santer was not amused. He felt that McIntyre was more interested in conducting "fishing expeditions" for his data than in doing genuine research. And he figured that if anyone's paper needed auditing it was not his, but Douglass's.

Santer's subsequent emails to his colleagues show his rising concern that he faced a return to the nightmare of 1996. On 11 November he told Karl, one of the 17 co-authors: "I believe McIntyre is pursuing a calculated strategy to divert my attention and focus away from research... I believe that our community should no longer tolerate the behaviour of Mr McIntyre and his cronies." He called McIntyre the "self-appointed Joe McCarthy of climate science", adding: "We should be able to conduct our scientific research without constant fear of an 'audit' by Steven McIntyre; without having to weigh every word we write in every email we send to our scientific colleagues."

Many would see references to Joe McCarthy as somewhat over the top. But the rich range of abuse thrown at Santer in the past means it is at least understandable. And of course the reference to weighing every word in emails was rather prescient. Santer's mood had not improved the following month, when he wrote an email to Wigley: "I'm damned and publicly vilified because I refused to provide McIntyre with the data he requested... Had I acceded, I am convinced I would have spent years of my scientific career deal-ing with demands for further explanations, additional data, Fortran codes [a programming language] etc... For the remainder of my scientific career I'd like to dictate my own research agenda."

But he was on difficult ground with his bosses. Around this time, he also wrote to colleagues: "After reading Steven McIn-tyre's discussion of our paper on climateaudit.com (and reading about my failure to provide McIntyre with the data he requested), an official at DOE [the US Department of Energy] headquarters has written to Cherry Murray at LLNL [the Lawrence Livermore National Laboratory], claiming that my behavior is bringing LLNL's good name into disrepute. I'm getting sick of this kind of stuff, and am tired of simply taking it on the chin."

But take it on the chin he did. After a further two weeks, he notified his co-authors that he had decided to publish online much of the data requested by McIntyre. He now reasoned: "This will make it difficult for McIntyre to continue making the bogus claim that he is being denied access to the climate model data necessary to evaluate the validity of our findings." Essentially he concluded that this was the path of least resistance, telling colleagues the following month: "I agreed to this publication process primarily because I want to spend the next few years of my career doing research. I have no desire to be 'taken out' as a scientist, and to be involved in years of litigation."

His change of mind brought a resounding slap on the back from Wigley, another co-author. Wigley said a spirit of openness would be beneficial all round. "Dear Ben," Wigley wrote a week before Christmas 2008. "This is a good idea. However will you give only tropical... results? I urge you to give data for other zones as well... To have these numbers online would be a great benefit to the community. In other words, although prompted by McIntyre's request, you will actually be giving something to everyone."

Wigley went on to suggest that this could be the start of a more general opening-up of Santer's data files. He asked: "What period will you cover? Although for our paper we only give data from 1979 onwards, to give data for the full 20th century runs would be of great benefit to all... This is a lot of work – but the benefits to the community should be truly immense."

CHAPTER 11

Silencing critics

*"This is truly awful. If we can find documentary
evidence of this, we could get him ousted"*

<small>TOM WIGLEY ON A TROUBLESOME JOURNAL EDITOR, JANUARY 2005</small>

Scientists sometimes like to portray what they do as divorced from
the everyday jealousies, rivalries and tribalism of human relation-
ships. What makes science special, they say, is that data and results
that can be replicated are what matters, and the scientific truth
will out in the end. An adversarial approach is part of that. So
friction between competing camps is to be expected, even encouraged. But a close reading of the emails exposes strenuous efforts
by the mainstream climate scientists to do what some outside
observers would regard as censoring their critics. And, for many,
the correspondence raises awkward questions about peer review –
one of the central pillars on which scientific credibility is built – as
well as about the operation of the IPCC.

The scientists involved deny they have been involved in
censorship. They say they are engaged not in suppressing dissent
but in upholding scientific standards by keeping bad science out
of peer-reviewed journals. Peer review, they say, exists precisely to
maintain the quality of the published literature. If it did not reject

some papers or force revisions in others, it would have no purpose. Gavin Schmidt at RealClimate goes further: "To have your opinion published in peer-reviewed literature is not some fundamental right – it is a privilege that depends on your ability to do the analysis and then marshal the logical arguments and data to support your point."

Under the peer review system, papers submitted to scientific journals are reviewed anonymously by experts in the field. Conducting reviews is seen as part of the job for academics, who are generally not paid for the work. The papers are normally sent back to the authors for improvement and only published when the reviewers give their approval. But the system relies on trust and usually on anonymity, so reviewers can give candid opinions. Editors are not bound to follow the advice of their reviewers, but as we show later, all hell can break loose when they do not.

A problem endemic in peer review is conflicts of interest that would not be allowed in most professions. Inevitably, papers will sometimes criticise the leading figures in the field who are judging them. The problem is not particular to climate science, of course. Cracks in the system have been obvious for years. In February 2010, 14 leading researchers in a different field – stem cell research – wrote an open letter to journal editors to highlight their dissatisfaction with the process. They alleged that a small scientific clique was using peer review to block papers from other researchers.

Of course, there can be value in asking scientists being criticised to review the charges against them. It can clear up misunderstandings or misinterpretations. The danger is that scientists being criticised thereby gain what amounts to a power of veto over criticism, especially if they are more eminent that their editors. The charge is vehemently denied by Jones and others, but many of those who have read the CRU emails have concluded that some scientists were wielding, or believed they wielded, just that power.

Phil Jones was regularly asked to review papers. His reviews were valued by editors. In 2006, the journal *Geophysical Research Letters* gave him an editors' award for his "conscientious and constructive reviewing". Nonetheless, he sometimes wrote critical reviews that might have had the effect of blackballing papers criticising his work. Here is how it worked in one case.

A key component in the story of 20th century warming is data from weather stations in sparsely populated regions such as Siberia. This huge area appears to have seen exceptional warming of up to 2C in the past century. But in such a remote region, actual data is thin on the ground. So how reliable is that data, and do scientists interpret it correctly?

In March 2004, Jones wrote to Mike Mann saying that he had "recently rejected two papers [one for the *Journal of Geophysical Research* and one for *Geophysical Research Letters*] from people saying CRU has it wrong over Siberia. Went to town in both reviews, hopefully successfully. If either appears I will be very surprised, but you never know with GRL." This email has been widely quoted. Understandably, given its extravagant language and Jones's evident confidence of success. What is notable is that Jones is nowhere talking about improving a paper or making its arguments fit for publication – merely of wielding his influence to keep out of print a paper criticising his work.

Jones did not specify which papers he had reviewed, nor what his grounds for rejecting them were. He has not clarified the matter since. But the *Guardian* has established that one was probably from Lars Kamel, a Swedish astrophysicist formerly of the University of Uppsala.

Kamel analysed the temperature records from weather stations in part of southern Siberia, around Lake Baikal. He claimed to find much less warming than Jones, despite analysing much the same data. Kamel told the *Guardian*: "Siberia is a test case, because it is supposed to be the land area with the most warming in the 20th century." It was a rare example of someone trying to replicate Jones's analysis. Kamel suggests the possibility of such discrepancies "should be checked in more regions and even globally". The finding sounded important. But his paper was rejected by *Geophysical Research Letters* that year. Kamel was leaving academic science and never tried to publish the paper elsewhere. But the draft seen by the *Guardian* asserts that the difference between his findings on Siberia temperatures and those of Jones is "probably because the CRU compilation contains too little correction for urban warming". Kamel says he no longer has a copy of the anonymous reviews of the paper, so we don't know why it was rejected. The paper could certainly be criticised for being slight, and for not revealing details about its methods of analysis. In which case, one approach for a reviewer would have been to press for further information rather than to reject the paper out of hand.

In an interview in February 2010, talking generally about his role as a reviewer, Jones said: "I do not accept that I was trying to subvert the peer review process and unfairly influence editors in their decisions. I undertook all the reviews I made in good faith and sent them back to the editors." Gavin Schmidt at RealClimate defended Jones: "It is not enough to simply call into question previous work. This is exactly the problem with the sceptics – the science needs to move forward, you have to do analysis. Otherwise the literature would simply be full of unsubstantiated claims which never get resolved."

But critics argue that this approach may, on the contrary, cause a proliferation of unresolved claims. McIntyre wrote on his website in December 2009: "CRU's policies of obstructing critical articles in the peer-reviewed literature and withholding data from critics have unfortunately placed issues into play that might otherwise have been settled long ago."

In another email exchange, Jones's deputy at CRU, Keith Briffa, initiates what looks like an attempt to have a paper rejected. In June 2003, as an editor of an unnamed journal, Briffa emailed fellow tree ring researcher Edward Cook, a researcher at the Lamont-Doherty Earth Observatory in New York, saying: "I have to nag you about that review. Confidentially I now need a hard and if required extensive case for rejecting [an unnamed paper] – to support Dave Stahle's and really as soon as you can. Please." Stahle is a tree ring professor from the University of Arkansas.

This request appears to subvert the convention that reviewers should be both independent and anonymous, though Schmidt says this is not necessarily so. "People often sign reviews and this could well have been a second go-around on a particular paper whose first-round reviews would have been seen by everyone concerned. Briffa (like many editors) can have a feeling that a paper should be rejected for multiple reasons but would like to have the reasons gone into in some detail."

Cook replied later that day: "OK, today. Promise. Now, something to ask from you. Actually somewhat important too." The favour was to provide some data to help Cook review a paper "written by a Korean guy and someone from Berkeley, that claims that the method of reconstruction that we use in dendroclimatology is wrong, biased, lousy, horrible, etc." He continued: "If published as is, this paper could really do some damage. It won't be easy to dismiss out of hand as the math appears to be correct theoretically, but it suffers from the classic problem of pointing out theoretical deficiencies, without showing that their improved [inverse regression] method is actually better in a practical sense."

The suspicion here is that Cook was intent on recommending rejection of a paper because it would "do some damage" rather than because it was poor science. Cook told the *Guardian*: "These emails are from a long time ago and the details are not terribly fresh in my mind." But Schmidt said: "The work in question has subsequently been published and the authors themselves have said that [it] was much improved over the initial submission."

Jones did not restrict his criticism of papers he saw as flawed to pre-publication reviews. He also had a reputation for harsh criticism of journals that regularly published papers he disagreed with. He defended this in an interview in February 2010, saying: "In some emails I questioned the peer review process with respect to what I believed were poor papers that had appeared. On some occasions I joined with others to submit a response to some of these papers. Isn't this called freedom of speech?"

Mann was also forthright in confronting journals he regarded as substandard. In March 2003, he discussed encouraging colleagues to boycott *Climate Research*, a small but influential journal published by Inter-Research, a company founded by German marine biologist Otto Kinne. Mann was angry about that journal's publication of a series of sceptical papers that, he said, "couldn't get published in a reputable journal". His anger had evidently been building for some time, but was focused in 2003 on a paper published in January that year and written by Harvard astrophysicists Willie (Wei-Hock) Soon and Sallie Baliunas. Soon, who was born in Malaysia, has long argued for the primacy of solar influences on climate change. Baliunas was previously known for contesting the idea that man-made chemicals were destroying the ozone layer.

The paper was an early attempted refutation of the hockey stick. After analysing 240 studies of past temperatures from tree rings and other sources, the authors said the proxies showed "that it was possible to identify a 50-year period in which temperatures were warmer than any 50-year period in the 20th century in most of the locations of the climate proxies". This is a carefully nuanced statement. It is not claiming that these 50-year periods all came together at the same time, thus constituting global warming of the kind seen in the 20th century. But it is saying nowhere is uniquely warm today.

Reasonably, the pair noted that "our results suggest a different interpretation of the multi-proxy climates compared to recent conclusions of Mann *et al.*" But for good measure they added the far more contentious claim that "the 20th century is probably not the warmest nor a uniquely extreme climatic period of the last millennium". The paper, and a later version published under the auspices of the George C Marshall Institute, had caveats sprinkled through it. Even the Marshall Institute version made clear that "the proxy data are far too incomplete to allow a realistic estimation of global surface temperature". But its political purpose was clearly to reopen a debate that the IPCC had sought to close two years before.

Harvard University press-released the paper under the headline "20th century climate not so hot" and continued without the caveats or a hint of a "probably", to state that "the 20th century is neither the warmest century nor the century with the most extreme weather of the past 1,000 years". The press release went on to say the paper showed that the medieval warm period and the little ice age were both "worldwide phenomena" and that "many parts of the world show the medieval warmth to be greater than that of the 20th century." None of this certainty was claimed in the original paper, though it would certainly have pleased lobbyists against the climate change consensus from the George C Marshall Institute, which later in the year described Soon and Baliunas as its "senior scientists", and the American Petroleum Institute, which the paper lists as helping to pay for the research, along with the Air Force Office of Scientific Research and NASA.

Mann emailed me at the time to say that the paper was "absurd, almost laughable (if it wasn't, as is transparently evident, being used as a policy- and politics-driven publicity stunt to support the dubious positions on climate change of some prominent American politicians)". He said, correctly, that Soon and Baliunas made no serious attempt to show whether the warmth they found at different places and times round the world in past eras were contemporaneous in the way current global warming is. If they were just one-off scattered warm events, they did not demonstrate any kind of warm era at all, just regional fluctuations in climate. Soon did not respond to *Guardian* requests to discuss the paper.

Mann clearly saw the study as a direct attack on his work. The emails show him undecided at first between ignoring the pair and attacking them. In March 2003, he told Jones: "I believed our only choice was to ignore this paper. They've already achieved what they wanted – the claim of a peer-reviewed paper. There is nothing we can do about that now, but the last thing we want to do is bring attention to this paper."

But Jones told Mann: "I think the sceptics will use this paper to their own ends and it will set [the field of paleoclimate research] back a number of years if it goes unchallenged. I will be emailing the journal to tell them I'm having nothing more to do with it until they rid themselves of this troublesome editor." The editor he had a problem with was Chris de Freitas of the University of Auckland in New Zealand, a man with known sceptical views about climate change who had worked on the paper. Jones continued: "A CRU person is on the editorial board [Clare Goodess], but papers get dealt with by the editor assigned by [editor-in-chief] Hans von Storch."

Jones was right about the political flak that would follow. The Marshall Institute version of the paper, published in June that year, was later read into the Senate record and taken up by the Bush administration, which attempted to get it cited in a report from the Environmental Protection Agency, apparently against the wishes of the report's authors.

Persuaded that the paper could not be ignored, Mann assembled a group of colleagues to pass judgment on it – effectively commissioning his own private peer review. The group included regular CRU emailers Phil Jones, Keith Briffa, Tom Wigley and Kevin Trenberth. They sent their findings to the journal's editorial board, arguing that Soon's study was little better than anecdote. They charged that it had cherry-picked data showing warm periods in different places over several centuries, followed by a globally cold little ice age, and had provided no evidence that demonstrated any overall warming of the kind seen in the 20th century.

While Soon and Baliunas had already got their publicity for the paper through their champions among media and lobby groups with a political agenda, Mann and his friends were determined to take on the journal that had published the original paper. They wanted nothing less than a retraction. And the emails reveal that when *Climate Research* failed to disown the paper, the scientists decided that what Mann called a "coup" had taken place, and that de Freitas was fast-tracking sceptical papers onto its pages.

Mann saw an irony in what had happened. "This was the danger of always criticising the sceptics for not publishing in the peer-reviewed literature. Obviously, they found a solution to that – take over a journal!"

Mann had a solution. "I think we have to stop considering *Climate Research* as a legitimate peer-reviewed journal. Perhaps we should encourage our colleagues… to no longer submit to, or cite papers in, this journal. We would also need to consider what we tell or request of our more reasonable colleagues who currently sit on the editorial board."

Was this improper pressure? Bloggers responding to the leaking of these emails believe so. But Gavin Schmidt at RealClimate denies it. Responding to our original report of these emails, he said: "No boycott was organised! Mike made a suggestion, for which you have no evidence for any follow-up. Though even if there had been an organised pressure campaign, why would that be improper? Voting with your 'feet' (or papers) is in fact something inevitable and is very sensitive to journal reputation without any actions on Mike's side whatsoever."

Mann himself denies wanting to "stifle legitimate sceptical views". He maintains that all the emails show is that he wanted to uphold scientific standards. "Please understand the context of this," he says. "This was in response to a very specific, particularly egregious incident in which one editor of the journal was letting in a paper that clearly did not meet the standards of quality for the journal."

When Mann presented his alternative peer review of Soon and Baliunas to the 10-man editorial board of *Climate Research*, however, a number took fright. They took Mann's side against their own editor. This was partly because they discovered that de Freitas had ignored the anonymous advice of four reviewers who wanted him to reject the paper. The board was in revolt. Their chief editor, von Storch, who had been with the journal for a decade, wrote an article saying the Soon paper shouldn't have appeared because of "severe methodological flaws". It amounted to the retraction Mann had demanded.

But the journal's publisher, the 80-year-old Otto Kinne, refused to publish von Storch's article without making revisions. He said: "Chris de Freitas has done a good and correct job." Mann responded angrily in emails to his colleagues, repeating his earlier entreaties to boycott the journal, saying: "I think that the community should… terminate its involvement with this journal at all levels – reviewing, editing, and submitting, and leave it to wither away into oblivion and disrepute."

That never happened, because von Storch and four other board members resigned and Kinne himself backtracked a few weeks later, admitting that publication had been an error and promising to strengthen the peer review process. He wrote in the journal that autumn that the "major conclusions" of the paper, specifically the claim that the 20th century is probably not the warmest nor a uniquely extreme climatic period of the last millennium "cannot be concluded convincingly from the evidence provided in the paper. CR should have requested appropriate revisions of the manuscript prior to publication… There was insufficient attention to the methodological basis of statements that touch on hotly debated controversies and involve pronounced political and economic interests."

It was a single sentence in the paper that went too far. But of course it was the sentence that mattered. Mann had won his argument. But the blood-letting involved in that victory ensured that this became a hugely divisive issue in which the personal often seemed more important than the scientific.

De Freitas has a long-standing public profile in both Canada and New Zealand as a climate contrarian. He was a driving force behind the creation of an early sceptic organisation in Canada, Friends of Science, which he set up with his brother Tim, a geologist with Imperial Oil, ExxonMobil's Canadian subsidiary. Stefan Rahmstorf of the Potsdam Institute for Climate Impact Research in Germany, a founder with Mann of the RealClimate website, contributed the following to the *Guardian* in 2010: "De Freitas had strong political views as seen by his numerous articles. For example, in an editorial in the *New Zealand Herald* on 14 January 2000, he compared climate scientists to Y2K bug consultants trying to cash in by scaring the public. And in 2002 he wrote in the *Bulletin of Canadian Petroleum Geology* that 'global temperature has not risen appreciably in the last 20 years'. A blatantly false statement."

Rahmstorf also clearly has strong views about de Freitas as an editor. He continued: "As editor [of *Climate Research*], de Freitas published several papers in line with his political views against clear advice from the scientific reviewers. Other members of the editorial board, such as Wolfgang Cramer [also of the Potsdam Institute], resigned in protest long before von Storch and the Soon paper affair. If scientists with such diverse views, after many discussions, all come to conclude that resigning is the only decent option, this clearly shows how much de Freitas had compromised the peer review process. No scientist would advocate keeping scientifically well-founded, differing viewpoints out of the peer-reviewed literature. But keeping politically motivated papers with flawed methodology out of the scientific record is the professional duty of all scientific reviewers. Those who became concerned by the peer review loophole that de Freitas had created can only be commended."

De Freitas declined to comment on these charges. But he defended his actions as an editor of *Climate Research*. "I was never ever found to have done anything wrong, even in rumpus over the Soon and Baliunas paper," he insisted to the *Guardian*. "All accusations against me were fully investigated and my performance as editor of this journal was shown to be flawless." This interpretation of events is supported by sceptical climatologist and Cato Institute fellow Pat Michaels. He alleged in the *Wall Street Journal* in December 2009 that the resignations of von Storch and his colleagues were a counter-coup initiated by Mann and Jones.

This is vehemently denied by von Storch. While one of the editors who resigned was a colleague of Jones at CRU, von Storch has a track record of independence and had been particularly critical of Mann's hockey stick. The emails reveal that far from being a closet supporter of Mann and company, he was widely and frequently reviled by them as "duplicitous" and worse. Yet writing in the *Wall Street Journal* the same month, von Storch said: "I am in the pocket of neither Exxon nor Greenpeace, and for this I come under fire from both sides – the sceptics and alarmists – who have fiercely opposing views but are otherwise siblings in their methods and contempt... I left the post [as chief editor of *Climate Research*] with no outside pressure, because of insufficient quality control on a bad paper – a sceptic's paper, at that."

A year after the publication of the paper by Soon and Baliunas, in July 2004, Jones wrote an email to Mann about two new papers. Both discussed an old issue – the influence of urbanisation on global warming, in particular whether it might be responsible for more warming than Jones and his colleagues believe. This dispute was of course what lay behind the "Wang affair" of the missing information on the location of Chinese weather stations. As then,

Jones appears, to put it mildly, impatient with anyone inclined to question his view.

One of the papers Jones objected to was by Eugenia Kalnay and Ming Cai of the University of Maryland. The pair had been arguing that urbanisation, combined with other changes in land use, was having an impact on global temperatures twice as great as previously thought. The other paper was identified in the email as being by "MM". This short form is common in the emails and often refers to Steve McIntyre and Ross McKitrick, the terrible twins coming to haunt the working lives of the emailers. But on this occasion it was almost certainly a reference to a paper from McKitrick and Pat Michaels in *Climate Research,* the journal at the centre of the Soon and Baliunas controversy the previous year. The paper too addressed the role of urbanisation, claiming to show a correlation between the geographical patterns of warming and of industrialisation and economic development. McKitrick said the correlations "are large and statistically significant", suggesting that urbanisation was a key element in warming.

Jones, embroiled once again in his pet-hate subject, called the paper "garbage". More damagingly, he added in an email to Mann with the subject line HIGHLY CONFIDENTIAL: "I can't see either of these papers being in the next IPCC report. Kevin [Trenberth] and I will keep them out somehow – even if we have to redefine what the peer review literature is!" This has, rightly, become one of the most famous of the emails. The bit about redefining peer review was no doubt a joke. But for once the overall intent seems to mean pretty much what it appears to mean. Jones wanted the papers kept out of the IPCC record if at all possible.

Jones and Trenberth had recently become joint lead authors for a key chapter in the next IPCC assessment report, known to insiders as AR4. They had considerable power over what went into those chapters, and to have ruled the papers out in such a manner would have been a clear abuse of the IPCC process.

Today, neither man attempts to deny that Jones's promise to keep the papers out was a serious error of judgment. Trenberth told the *Guardian*: "I had no role in this whatsoever. I did not make and was not complicit in that statement of Phil's. I am a veteran of three other IPCC assessments. I am well aware that we do not keep any papers out, and none were kept out. We assessed everything [although] we cannot possibly refer to all literature... Both of the papers referred to were in fact cited and discussed in the IPCC."

In an additional statement agreed with Jones, he said: "AR4 was the first time Jones was on the writing team of an IPCC assessment. The comment was naive and sent before he understood the process." Some will not be content with that. AR4 was indeed the first assessment report in which Jones had been a lead author, responsible for the content of a whole chapter. But he had been a contributing author to such reports for more than a decade and should have been well aware of the rules.

Trenberth is right to say that, in the end, the McKitrick and Michaels paper was discussed in the IPCC report. But its passage was not simple. McKitrick points out that it was not mentioned in either of the first two chapter drafts sent to reviewers. Mention of it was only added in the final draft of the report, which would not have been shown to reviewers. So the IPCC's treatment of the paper would not have been widely discussed. Given that it was implicitly criticising the work of chapter lead author Jones, and that we now know Jones had initially wanted all reference to it expunged, this is troubling. And so is the nature of the discussion of M&M's paper that made it into the report.

The final, published version of the IPCC chapter notes the link between areas of greatest warming and areas with urbanisation, as reported by McKitrick and Michaels. But it goes on to dismiss the link as pure chance. It argues that these areas of urbanisation were "also those that have been most warmed by atmospheric circulation changes... Hence the correlation with industrial and socioeconomic development ceases to be statistically significant." Oddly, this dismissal of the original paper is not sourced – even though it has a precise mathematical meaning that could be tested. McKitrick said in 2010 that the IPCC statement remains unverified in published literature.

Climate Research is a fairly minor journal. Not so *Geophysical Research Letters*, published by the august American Geophysical Union (AGU). But when it began publishing what Mann, Wigley, Jones and others regarded as poor-quality sceptical papers, the emailers again responded angrily in their discussions.

As we have seen, in early 2005 *Geophysical Research Letters* provided a home for the paper by McIntyre and McKitrick that made the most damaging charges about the statistical methods used in Mann's hockey stick analysis. Afterwards, Mann complained to the journal. Its editor-in-chief, Steve Mackwell, replied patiently: "I have reviewed the manuscript by McIntyre, as well as the reviews. The editor in this case was James Saiers. He did note initially that the manuscript did challenge published work, and so felt the need for an extensive and thorough review. For that reason, he requested reviews from three knowledgeable scientists. All three reviews recommended publication. While I do agree that this manuscript does challenge (somewhat aggressively) some of your past work, I do not feel that it takes a particularly harsh tone."

Mackwell suggested Mann might like to write a comment by way of reply. Mann was not satisfied. He saw not a disagreement over a paper, but a new battle in an ongoing war. He wrote to colleagues later that day: "Apparently the contrarians now have an 'in' with GRL. I'm not sure that GRL can be seen as an honest broker in these debates any more."

Gavin Schmidt at RealClimate says today that Mann's interpretation was reasonable. A "systematic" problem had arisen at GRL "over the period 2005–2006", he says. "There was a string of bad papers published – ones that did not properly support their conclusions and made basic errors in the science." In this connection, he listed five specific papers that had come to the attention of Mann and others. The authors are familiar to anyone who follows the literature from the climate sceptics. Four of the papers carried the name of David Douglass, who featured earlier in our story. Two involved the legendary proponent on solar influence of climate change, Fred Singer, and one involved Pat Michaels. "Are scientists not supposed to notice these patterns?" Schmidt asks. "Or never discuss them among colleagues?"

Noting patterns, of course, is one thing. But the emails reveal that Mann had checked out the editor James Saiers. He was a chemical engineer at Yale. He told colleagues that Saiers had a "prior connection" with the same department at the University of Virginia where sceptic Pat Michaels worked. He added: "We now know" how various other sceptically tinged papers had got into GRL. Wigley, one of the recipients of that email, appeared to agree. "This is truly awful," he said, adding that "if you think that Saiers is in the greenhouse sceptics' camp, then, if we can find documentary evidence of this, we could go through official AGU channels to get him ousted." This doesn't appear to have happened, but it would have been a startling politicisation of the process of journal editing.

In 2006, a year after Mann and co noticed the "in" that sceptics appeared to have gained at GRL, and had linked it to Saiers's role as an editor, Saiers gave up the GRL post. Sceptics have claimed that this was due to pressure from Wigley, Mann and others. Saiers says his three-year term was up. "My departure had nothing to do with attempts by Wigley or anyone else to have me sacked," he told the *Guardian*. "Nor was I censured, as I have seen suggested on a blog posting written by McKitrick."

As for Mann's smear about a link between him and Michaels, Saiers says he does not remember ever talking to Michaels while they were both at the University of Virginia, "though I did attend a barbecue at his home back in the early 1990s." He concludes that "Wigley and Mann were too keen to conclude that I was in league with the climate change skeptics. This kerfuffle could have been avoided if the parties involved would have done more to control their imaginations." Not, perhaps, the only time.

CHAPTER 12
Free the data

*"Why should I make the data available to you,
when your aim is to try and find something wrong with it?"*

PHIL JONES TO WARWICK HUGHES, FEBRUARY 2005

The leaked emails reveal repeated attempts by Phil Jones and his colleagues at CRU to block freedom of information requests from climate sceptics and the plain curious who wanted access to correspondence, documents and data. Not only do these moves appear contrary to the spirit of scientific openness, but according to the Information Commissioner's Office (the British government body that administers the FoI legislation), the requests were on at least one occasion "not dealt with as they should have been under the legislation".

Jones says he did things by the book. He told an interviewer in February 2010 that he had always followed the university's policy and guidelines on information law, and that the university's information policy and compliance manager (IPCM) took responsibility for co-ordinating responses to requests. "I worked with those colleagues and the IPCM to handle the requests, with responses going from the IPCM. He also liaises with the Information Commissioner's Office where necessary and did so on several occasions in relation to requests made to CRU."

But the law about how freedom of information should apply to scientists is largely untested in the courts. And the emails reveal deep and understandable frustration among the scientists about the time and energy they were being asked to give up to dealing with the requests. This was particularly galling as the sceptics making the requests were, in the scientists' eyes, not interested in advancing human knowledge, but wanted to disrupt their work, fish for information and pick holes in their analyses to suit an anti-global warming agenda.

Freedom of information law is a recent innovation in Britain. It was enacted in 2000. Few scientists initially realised that it might impinge on their work. But one who did was Jones. Thanks to his brushes with climate sceptics, he could see that the laws would put new powers into their hands. And he foresaw early on that his arch-inquisitor Steve McIntyre would be a thorn in his side.

In 2005, the emails show that Jones was both laying out his uncompromising views on protecting "his" data from FoI snoopers and battening down the hatches against data thieves. In a note to Mike Mann in February that year, he said: "Just sent loads of station data to Scott." This was Scott Rutherford, a climate scientist at the Roger Williams University, Rhode Island, who was then collaborating with Mann on a research paper using CRU data. Rutherford had recently left some data and computer code on an open FTP (file transfer protocol) site so that Tim Osborn at CRU could access it. Instead, McIntyre discovered it.

Jones continued: "Make sure he documents everything better this time! The two MMs [McIntyre and McKitrick] have been after the CRU station data for years. If they ever hear there is a Freedom of Information Act now in the UK, I think I'll delete the file rather than send to anyone." He added: "We also have a Data Protection Act, which I will hide behind."

Later, in 2007, Jones told Wei-Chyung Wang, his co-author on the notorious 1990 paper discussing the Chinese urban heat island, and Tom Karl, director of the US government's National Climate Data Centre: "Think I've managed to persuade UEA [the University of East Anglia] to ignore all further FoI requests if the people have anything to do with ClimateAudit."

In December 2008, Jones elaborated in an email to Ben Santer how he had achieved this agreement. "When the FoI requests began here, the FoI person said we had to abide by the requests. It took a couple of half-hour sessions – one at a screen – to convince them otherwise, showing them what CA [ClimateAudit] was all about. Once they became aware of the type of people we were dealing with, everyone at UEA... became very supportive. I've got to know the FoI person quite well and the chief librarian – who deals with appeals. The VC [vice-chancellor] is also aware of what is going on." This puts into context his insistence that he was working by the book.

By and large, the records show that the university turned down most FoI requests, from McIntyre and others, for CRU data. An exception was a successful FoI request by the *Guardian*'s technology editor, Charles Arthur, submitted in December 2009 and answered the following month. It uncovered the fact that of 105 requests concerning CRU submitted up to December 2009, the university had by late January 2010 refused 77, accepted 6 in part, had 11 outstanding, and had acceded in full to only 10. One was withdrawn.

By 2008, the CRU scientists had become used to dealing with, and usually rebuffing, requests for data. But in May that year, Jones received an FoI request from David Holland, an electrical engineer from Northampton. It asked for all emails sent and received by tree ring specialist Keith Briffa in relation to the IPCC fourth assessment of climate science (AR4) published the year before. Briffa, as we saw earlier, was a lead author for the report's chapter on paleoclimatology.

In response to past controversies, the IPCC had for the first time archived its formal review exchanges and put that material online. Anyone can read it. But Holland wanted to see email discussions between scientists about IPCC texts that he believed were being conducted outside that formal process.

He wasn't far wrong. It had become common practice to exchange papers and comments directly from scientist to scientist. The emails leaked the following year discussed this. A few days after Holland's application, Jones noted in an emails to CRU colleague Tim Osborn, copied to Briffa: "What we did get were papers sent to us directly – so not through IPCC, asking us to refer to them in the IPCC chapters. If only Holland knew how the process really worked!!" And they discussed getting their story straight. "Keith should say that he didn't get anything extra that wasn't in the IPCC comments."

Concern was focused on having to reveal email discussions about how the assessment would handle the troublesome Wahl and Ammann paper defending the hockey stick. These back-channel communications between the paper's authors and IPCC authors, including early versions of the paper, seemed a direct subversion of the spirit of openness intended when the IPCC decided to put its internal reviews online.

One line of defence against releasing the emails was to claim that they were confidential. Osborn wrote to Caspar Ammann, asking "whether you consider emails that you sent to us to have been sent in confidence. If you do, then we will use this as a reason to decline the request." Ammann's reply suggests he was embarrassed and confused by the question. "As far as I can remember... I have never written an explicit statement on these messages that

would label them strictly confidential." Nonetheless, along with
the cost of compliance, confidentiality was ultimately the argument that the University of East Anglia used to refuse Holland's
request for IPCC emails.

In the midst of this, Jones sent one of the most damaging
emails to emerge from the leak. He asked Mann: "Can you delete
any emails you may have had with Keith [Briffa] re AR4? Keith will
do likewise. Can you also email Gene [Eugene Wahl] and get him
to do the same... We will be getting Caspar [Ammann] to do the
same." This email seems to have been what persuaded the UK's
Information Commissioner's office that the university's handling
of FoI requests was not correct. The deputy information commissioner Graham Smith put out a statement in late January 2010
that said: "The emails which are now public reveal that Mr
Holland's requests under the Freedom of Information Act were
not dealt with as they should have been under the legislation.
Section 77 of the Freedom of Information Act makes it an offence
for public authorities to act so as to prevent intentionally the
disclosure of requested information."

Smith said the only reason the ICO would not be taking
action over the apparent breach was the statute of limitations. The
presumed offence occurred more than six months before. Jones
now says that the damaging email to Mann was "sent out of frustration", and that in fact no emails were deleted as a result.

There were more emails sent at around that time discussing
the deletion of emails. Jones also wrote to Ray Bradley at the
University of Massachusetts, saying: "You can delete this attach-
ment [Holland's FoI request] if you want. Keep this quiet also,
but this is the person who is putting FoI requests for all emails
Keith and Tim have written and received re Ch 6 of AR4. We
think we've found a way around this."

Another argument used to rebuff FoI requests for IPCC-related correspondence was that IPCC documents were not covered by British law. It seems that the Information Commissioner may at some stage have agreed to this interpretation. Jones wrote to NASA climatologist Gavin Schmidt in August 2008: "All our FoI officers have been in discussions and are now using the same exceptions not to respond – advice they got from the information commissioner... The FoI line we're all using is this. IPCC is exempt from any country's FoI – the sceptics have been told this. Even though we possibly hold relevant info, the IPCC is not part of our remit (mission statement, aims etc) therefore we don't have an obligation to pass it on."

Behind this legal hair-splitting, the emailers took a simple view that, whatever the status of data and whatever the law might say, personal emails should be sacrosanct. As Briffa told Ammann a month later: "I've been of the opinion right from the start of the FOI requests that our private inter-collegial discussion is just that – PRIVATE... submitting to these demands undermines the wider scientific expectation of personal confidentiality. It is for this reason, and not because we have or have not got anything to hide, that I believe none of us should submit to these 'requests'."

The anger persisted. Seven months later, in December 2008, Jones revealed in an email to Santer discussing McIntyre: "If he pays ten pounds (which he hasn't yet) I am supposed to go through my emails and he can get anything I've written about him. About two months ago I deleted loads of emails, so have very little – if anything at all." This remark was in connection with privacy legislation under the Data Protection Act, rather than freedom of information law. Jones has been quoted elsewhere as saying: "We haven't deleted any emails. I delete my own personal emails a year at a time regardless of subject as I have too many, but the university still has the emails."

Indeed so, as it transpired. He didn't know that a year later many of them would be all over the internet.

The University of East Anglia now says that no emails were deleted after these exchanges. But Holland, who had requested the IPCC emails, says such correspondence reveals "a deliberate attempt to destroy info which has been properly requested".

It was embarrassing, and perhaps demeaning, to have outsiders ask for your emails. But in the end it was a sideshow for most scientists compared with their data. Data was the lifeblood of their work, the stuff they sweated years to generate and analyse. And nobody felt that more acutely than the CRU staff assembling temperature data from round the world over decades. They did not take kindly to outsiders turning up and asking to be handed their crown jewels.

And so it was when one of Jones's earliest detractors, Warwick Hughes, came calling. Today Hughes describes himself as a free-lance earth scientist from Perth in Australia, and calls global warming a "fraudulent notion". Back in 1991, when he and Jones first crossed swords, Hughes was working for the Tasman Institute, a now defunct free-market thinktank based in Melbourne. He analysed a study of temperature trends in the southern hemisphere published by Jones five years before, and claimed that virtually all the warming found by Jones was a result of growing urban influences. Later, he investigated Jones's South African and Siberian data, claiming of the latter that "cities are the source of the apparent warming, which is not apparent at nearby small town or rural stations".

The leaked emails reveal a civilised correspondence in 2000 between Hughes and Jones. Jones admitted that Hughes had seemingly found significant "anomalies" in his published data and asked for more details about what he had uncovered. But in 2004, when Hughes asked Jones for monthly temperature data from 3,000 weather stations, described on the CRU website as "the foundation of Dr Jones's published papers", relations soured. This was, incidentally, the same period during which Jones's previously cordial exchanges with McIntyre deteriorated.

After six months of delay, Jones told Hughes in February 2005 that some of the data was confidential, having been obtained from the World Meteorological Organisation. But "even if WMO agrees, we will still not pass on the data. We have 25 or so years invested in the work. Why should I make the data available to you, when your aim is to try and find something wrong with it?"

This was at least honest. The statement is damaging nonetheless, because the entire purpose of scientific replication is to try to find something wrong with existing data and theories. That is how science advances. In emails to his colleagues, Jones often said confidentiality agreements were a useful excuse. As he told Wigley in January 2005: "Data is covered by all the agreements we sign with people, so I will be hiding behind them." But for Hughes, he didn't bother with the pretence.

During 2008 the emailers continued to debate what to do about a rising tide of FoI requests. Most saw them as a threat to their work – not because they would uncover fraud, but because they took up their time. Many thought disruption was the true purpose of the people who submitted the requests.

Schmidt wrote to Ben Santer in December 2008 about the latter's travails in dealing with McIntyre: "There are two very different things going on here. One is technical and related to the actual science... The second is political. The second is the issue here... Whatever you say, it will still be presented as you hiding data. The contrarians have found that there is actually no limit to what they can ask people for (raw data, intermediate steps, additional calculations, residuals, sensitivity calculations, all the code, a workable version of the code on any platform etc) and like Somali pirates they have found that once someone has paid up, they can always shake them down again."

Some took the line that their only option was to pull up the drawbridge. But others wanted to give some ground, whether in the name of good science or a peaceful life. They included the veteran climate scientist Stephen Schneider. Now at Stanford University in California, he is also founder and editor of the journal *Climate Change* and recent author of *Science as a Contact Sport*.

In January 2009, Schneider wrote a round-robin to scientists in which he agreed: "This continuing pattern of harassment... in the name of due diligence is in my view an attempt to create a fishing expedition to find minor glitches or unexplained bits of code – which exist in nearly all our kinds of complex work – and then assert that the entire result is thus suspect." Nonetheless, he argued that researchers should give enough data and information on their sources and methods so that those "who are scientifically capable can do their own brand of replication work".

He did not, however, feel that such openness should extend to "personal computer codes with all their undocumented subroutines etc... It would be [an] odious requirement to have scientists document every line of [computer] code so outsiders could just apply them instantly. Not only is this an intellectual property issue, but it would dramatically reduce our productivity since we are not in the business of producing software products for general consumption and have no resources to do so." Presciently, he added: "Good luck with this, and expect more of it as we get closer to international climate policy actions. We are witnessing the 'contrarian Battle of the Bulge' now and expect that all weapons will be used."

As the scientists resisted, anger grew among their critics. At the end of August 2009, an amateur British sceptic called Rupert Wyndham spotted that Phil Jones had recently been made a fellow of the American Geophysical Union, which published many of his papers and employed him as a reviewer of others. The citation mentions his diligence on reviewing papers for the union's journals, including the *Journal of Geophysical Research* and *Geophysical Research Letters*. It came two years after he received the editors' award from the latter journal.

Wyndham assembled an international group of sceptics from 10 countries and wrote to the AGU's top atmospheric scientist, Alan Robock, to complain. Signatories included Sonja Boehmer-Christiansen, editor of *Energy and Environment*, and Martin Durkin, the British TV producer briefly notorious for his programme *The Great Global Warming Swindle*. They accused Jones of a range of data crimes. "Honouring a man who consistently breaches the fundamental protocols of scientific method casts a stain on the reputation of the AGU," they wrote.

In retrospect, it was clear that by 2009 things were coming to a head. Freedom of information requests were piling up. The scientists were increasingly angered at how long it was taking to fend them off. It would, they felt, be even worse to deliver all the data being requested. They could only see that such acquiescence would result in further demands as the "Somali pirates" dug their teeth into the data.

Hughes's demands for CRU's crown jewels had receded into the background. But his place had been taken by McIntyre, who showed much greater perseverance and tenacity. Stories began to circulate in the science community about how CRU was resisting legitimate requests from McIntyre. In early July 2009, when I asked Jones about these rumours, he told me: "McIntyre has no interest in deriving his own global temperature series. He just wants to pick holes in those who do. He wants not only the original station data, but details of all the adjustments we have made over the years. It's just time-wasting."

McIntyre had been demanding, through freedom of information requests, the raw temperature data from which Jones had assembled his graphs of global temperature trends over the previous 160 years. In June, McIntyre had noticed a note on the Met Office's website saying that raw data from the joint Met Office and CRU database was available. He asked Jones for the data. He did not want the processed data, which had already been massaged to correct for malign influences like urbanisation and weighted according to the area of the earth's surfaces that the data was taken to represent. He wanted the raw data. Like Hughes before him, he wanted to see how the data massaging was done.

The data set was huge. By that time it comprised approaching more than 4m pieces of temperature data from more than 4,000 weather stations. Much of the data was used by one of the two other main centres analysing global temperatures, at NASA's Goddard Institute for Space Studies and NOAA's National Climate Data Centre. NASA had a much more open policy on releasing that data to outsiders. Much of the rest was in the public domain, held at the World Meteorological Organisation and elsewhere. But much was not. And without the complete set, replication of CRU's work was not possible.

Nonetheless, Jones rejected McIntyre's request for this data, on the grounds that the information was not his to give. The data did not come from his thermometers, after all. And much of it had been obtained under confidentiality agreements from the national meteorological services that supplied it. McIntyre was unimpressed. He noticed that Jones did share the data with fellow academics who, in the normal course of scientific publication, would be expected to put Jones's name on their papers as a co-author. McIntyre said that he wanted the same data that Jones had recently given freely to Peter Webster, a climate scientist at the Georgia Institute of Technology in the US. Still no dice.

The day after the formal rejection of his demand for this data, McIntyre announced that he had a full set of the station data. He headlined his short article "A mole". And said: "Folks, guess what. I'm now in possession of a CRU version giving data for every station in their station list." He published some, the temperatures from Lund in Sweden between 1753 and 1773 – "sensitive information indeed", he noted sarcastically on ClimateAudit.

Who was the mole? The comments on his website were agog. But the next day, he announced the mole had been found. Well, not quite. He said that "late yesterday I learned that the Met Office/CRU had identified the mole. They are now aware that there has in fact been a breach of security. They have confirmed that I am in fact in possession of CRU temperature data…" He did not say who his source for this information was.

Later McIntyre clarified his cryptic remarks. There was no mole. He had simply found the material on CRU's website. According to a subsequent article in *Nature*, McIntyre had stumbled on FTP files containing station data that was intended to be shared only with CRU's partners at the Met Office. McIntyre himself later said: "I downloaded from the public CRU FTP site… No hacking was involved." The "mole", he now says, was simply the person who "put the station data on the CRU server". He suggests that that person was Jones himself.

Some see this episode as the first step towards the leak of emails four months later. Certainly the emails were ultimately taken from just such an FTP server. But there is no specific evidence to suggest the events are otherwise linked. In any event, CRU immediately said it had removed the data from its website, leading to charges from McIntyre that they were engaged in a "purge".

With the data that he requested now apparently in his possession, McIntyre continued demanding to know why it could not have been handed out freely in the first place. He requested that CRU publish the confidentiality agreements that forced them to keep the data out of public sight. He found an interesting route to achieve that.

According to *Nature*'s blogger Olive Heffernan, "between 24 and 29 July, CRU received 58 FoI requests from McIntyre and people affiliated with ClimateAudit... The Met Office, which receives a cleaned-up version of the raw data from CRU, has received 10 requests of its own." The university later complained that this "unprecedented and frankly administratively overwhelming" rush of requests was intended to harass it. In fact it was a bit more complicated than that.

McIntyre says that he personally submitted only two FoI requests to CRU during 2009. One was for the CRU station data. The second, when he was told he could not have the data because it was held in confidence, was a request for copies of confidentiality agreements. He imagined there were dozens. He says he believed that, on past record, if he simply asked for all confidentiality agreements he would be denied. So he asked for five, covering Canada, the US, Australia, Britain and Brazil. And he called on his readers on ClimateAudit to ask for the rest, in batches of five countries each. He put a template request form on his website, and his followers did the rest, with fellow blogger Steve Mosher keeping a running check on which countries had been tagged.

The university has called this blizzard of requests vexatious. The word "vexatious" matters because it is a specific justification under the British legislation for refusing requests. McIntyre denies that his requests were vexatious or frivolous. He says that if Jones and his colleagues had either supplied the data or published the agreements earlier, then the rush of FoI requests would never have been necessary. In fact, Jones and his colleagues had become so intransigent that an FoI request was the only means of extracting data from them.

In any case, he points out, it was only the blizzard of requests that summer that flushed out the fact that the university could only put its hands on a handful of actual written agreements – from Norway, Bahrain, Spain and Britain's own Met Office. It said that the rest were either informal or lost.

In early July, McIntyre appealed to the University of East Anglia against being refused the CRU's raw temperature data. He might already have it, but there was a principle at stake. His appeal was turned down by the university's director of information services, Jonathan Colam-French, in a letter dated 13 November. McIntyre says he received that letter on the 18th. But by then, something else had happened. Climategate had happened. More than 1,000 emails, plus documents and data had been liberated/hacked/stolen from an FTP server on the CRU website.

McIntyre says the sequence of events suggests an inside job. The first attempt to put the zip file containing the CRU emails online happened on the morning of Tuesday 17 – after the letter had been written but before he had received it. The files contained emails sent or received up to the 12th, the day before the letter was sent. The circumstantial evidence, he concludes, suggests an aggrieved employee or student angry about the secrecy over CRU's data. Perhaps the data and emails were collected in a file for the purposes of deciding McIntyre's application. It was a large file, but after all this was CRU's crown jewels. Interestingly, as blogger Steve Mosher has pointed out, Colam-French's letter mentions that he consulted a file. Maybe there was a mole after all. We will return to the mystery of who copied and released the files later.

Whatever the truth about who carried out the hack, there is an irony buried in Jones's 2005 correspondence with Mann over keeping data secure from prying eyes. "Don't leave stuff lying around on FTP sites – you never know who is trawling them," he wrote. The tragedy for Jones is that his emails frequently reveal a prescient eye for the disaster that was destined to unfold around him.

PART THREE

A perfect storm

CHAPTER 13

The search for the hacker

"These three sentences could only have been written by a well-educated native speaker of English, probably a Britisher"

PROFESSIONAL TRANSLATOR, MARCH 2010

Who did the deed? Who hacked/liberated/discovered the emails? The answer would not change what they reveal about the writers. But it is the conundrum that everyone involved would love to solve. In late November 2009, the University of East Anglia called the Norwich police station to report a theft. A theft of data. Within days, Superintendent Julian Gregory had decided this was no task for local officers to conduct alone. He called in both the counter-terrorism squad and Scotland Yard's electronic crimes unit. But months later, as this book goes to press, it remains unclear even if a crime has been committed.

There were three stages to the publication of the large zip file, tagged "FOA2009.zip", on the internet. And each may have been done by someone different. First, there was the assembly of the material: 4,660 files, including documents, raw data and computer code. Some of the data, for instance on tree rings, dates back to 1991. The 1,073 files containing emails (often several in a string) began in 1996 and ended on 12 November 2009. This can only be a small subset of the emails sent and received by CRU staff during that time, so someone chose them.

The subset of emails is largely restricted to messages discussing work. There are no social memos or invitations to eat birthday cake in the lab. They cover many scientific issues, mostly without rancour or hint of conspiracy. Most involve just four individuals as senders or recipients. Three are CRU researchers: Phil Jones, Keith Briffa and Tim Osborn. The fourth is Mike Hulme, formerly of CRU but now director of the neighbouring Tyndall Centre for Climate Change Research. Of the 66 emails that do not involve these four people, most came via mailing lists to which at least one of the four belonged. Finally, most of the emails are between these four and foreign collaborators, particularly in the US.

At first sight it looks as if someone, probably with some knowledge of CRU's work, and perhaps from within CRU itself, collected the files for some unknown and perhaps quite innocent purpose. The most obvious would be that the file was assembled to help assess one of Steve McIntyre's freedom of information requests. Alternatively, the subset may be a result of some fairly crude sifting of a larger tranche of CRU emails using a search routine to pick out key words and phrases like Yamal and tree rings. This could have been done either before or after the material was copied.

Some have suggested that the emails could have been assembled only by trawling a wide variety of machines across the university, meaning only an insider could have done it. But the university says that all the material was sitting in an archive on a single back-up CRU server when it was copied.

The next step was copying the material. It was not copied all in one go. An analysis by the *Guardian*'s technology editor Charles Arthur revealed that the copying appears to have taken place over a number of weeks, mostly on three dates: 30 September, 10 October and 16 November 2009. But it remains unclear if this followed a security breach or whether someone leaked the material from within the system or deliberately or inadvertently left it somewhere where an outsider could copy it from a part of the CRU server available to outsiders. It may be relevant that all the copying apparently took place in the final weeks of consideration by the university's information officer of McIntyre's request for the CRU's crown jewels, and that the last addition to the file of leaked material was made the day before the decision letter was written.

Finally there was the distribution. We know a CD of the files existed prior to their widespread release, but also that they were loaded remotely onto websites. This was done in a reasonably sophisticated manner, using one of the "open proxies" favoured by hackers to cover their traces. At various points, those involved used servers in Turkey, Russia, Malaysia and Saudi Arabia. The location of these servers is probably entirely opportunistic. Hackers can go online and choose from a range of open proxies round the world. More interestingly, digital fingerprints on the files that were distributed show that they were created on a computer set variously at four and five hours behind GMT – placing the hacker on the eastern seaboard of North America, most likely the US or Canada.

Who might have been involved? Three groups of people have been suggested. They could have been dissidents from the University of East Anglia, perhaps angry with CRU scientists or the actions of the university's information office, which had lately been rejecting an unusual number of FoI requests to CRU. Such insiders might have had routine access to internal servers. People in the environment school, within which CRU was located, told the *Guardian* there had been some grumblings and jealousies about Jones and the CRU scientists, but no outright hostility.

There is no hint in the emails that the university information staff were anything other than friends to Jones. The senior staff often appeared to be too close. Nor do the emails contain any concerns about leaks from that office. But it is possible that an individual privy to decisions being taken about freedom of information requests may have agreed with the later comments from the national Information Commissioner's office that the university was not meeting the requirements of the law in releasing CRU data.

A second possible group is people acting for a corporation or even a state entity. They might have been anxious to influence either the climate negotiations being held in Copenhagen the following month, or internal US debates on climate. Here, the main evidence is the sophistication of the hack and release, leaving no known traces. And the timing. While Climategate did not have a direct effect on the Copenhagen negotiations, its timing just before that event ensured maximum publicity. It was also well timed to influence discussions in the US Senate on a climate change bill. This theory would be consistent with the known corporate "stealth" tactic of using citizens' groups to spearhead opposition to policies such as healthcare reform. And US lobbyists have a history of garnering publicity hostile to climate science before major climate negotiations. But I have seen nothing specifically linking corporate America to the hack.

Thirdly, there are the bloggers hostile to CRU's brand of climate change science. Besides motive, they would likely have had the internet skills to carry out a successful hack. And of course, their websites swiftly became the route through which the world discovered about the emails.

The first releases of the emails all involved a group of bloggers on America's west coast. Foremost was Steve Mosher, a statistician with declared libertarian instincts. He worked as a software engineer writing open-source software until June 2009. After that, and following previous collaboration with Steve McIntyre, he spent his time researching and blogging on climate change. Mosher had a clear motive. He says his efforts to do quality checks on CRU data were "thwarted by CRU's unwillingness to share the data as they used it". If anyone broke the Climategate story in news terms, it was him.

Weeks after the release of the emails, Mosher published an instant book, *Climategate: The Crutape Letters*, with co-author Thomas W Fuller. In a brief biography attached to his submissions to the House of Commons Climategate inquiry, Mosher said that in early 2010 this was his sole source of income. The book avoids denialism or hyperbolic rhetoric. "We do not for one minute believe there is any evidence of a long-term conspiracy to defraud the public about global warming," the authors say. "What we find evidence of on a much smaller scale is a small group of scientists too close to each other, protecting themselves and their careers, and unintentionally having a dramatic effect on a global debate."

Other key figures among the west coast bloggers included Anthony Watts, the San Francisco radio station meteorologist who runs the blog site Watts Up With That. The site began as a citizens' exercise to check the location of US weather stations to see if they were influenced by undeclared urban influences. But it has turned into a notice board for climate sceptics. With more than two million unique visitors a month, WUWT may be the most-read climate blog in the world. Watts runs it with his online moderator, Charles Rotter, Mosher's San Francisco flatmate. Rotter is known online as "Charles the Moderator". Also active, and early receivers of the file, were mechanical engineer Lucia Liljegren's blog The Blackboard, Jeff "id" Condon's The Air Vent, and Warren Meyer's Climate Skeptic blog out of Phoenix,

Arizona. None, however, is based on the east coast of the US, from where the *Guardian*'s analysis reveals the files were probably originally copied. The only leading blogger whose fingerprint matches that time zone is Steve McIntyre.

Here is what is known about how Climategate happened. During the autumn of 2009, someone was secretly copying files from a back-up server at the university's Climatic Research Unit. On 17 November at 6.20am EST, the day after the final email was released, someone tried to upload a zip file, FOI2009.zip, containing the CRU emails and other documents and data onto the climate scientists' own favourite website, RealClimate. This was done via a Turkish server, and apparently using a password. That person announced this with a post that read: "We feel the climate science is, in the current situation, too important to be kept under wraps. We hereby release a random selection of correspondence, codes and documents. Hopefully, it will give some insight into the science and the people behind it." The post gave 20 samples from the emails and a link to download more.

This initial message, incidentally, contains interesting clues to the possible identity of the hacker. A professional translator with extensive experience of working outside Britain pointed out to the *Guardian*: "These three sentences could only have been written by a well-educated native speaker of English, probably a Britisher. In particular, the absolutely correct usage of 'hereby' and 'hopefully' (in a particularly British-English manner) is conclusive. Note also that the first occurrence of 'it' is absolutely correct, although many people might have written 'they'. The use of the word order of the first sentence… is also indicative of high-level native speaker fluency in the written language. No Russian or Chinese or Turk or Saudi person could have written this passage."

Gavin Schmidt, the NASA scientist running RealClimate, swiftly spotted the rogue post and removed it from the site. Having read the files he alerted CRU. But even as he did that, a cryptic comment appeared on McIntyre's ClimateAudit site at 7.24am. "A miracle has happened," it said, providing a link to the post on the RealClimate website. Nobody noticed this initially. And only four people tried to activate the link before McIntyre eventually removed it.

Later that day in California, a link to a Russian server holding the zip file was posted to WUWT, where Charles the Moderator held it, alerted his boss Anthony Watts, ran virus checks and awaited approval to put it on the site. Watts was in Europe at the time. By the evening of the 17th, links to the zip file had also been posted to Jeff id's Air Vent blog and Warren Meyer's Climate Skeptic site, both unmoderated. San Francisco online journalist Patrick Courrielche, who has investigated the affair, says Jeff id, an aeronautical engineer, was out deer-hunting and didn't notice the upload till he got an email later from Mosher pointing it out.

McIntyre says that Charles the Moderator made backup CD copies of the file, while waiting for approval from his boss to upload the link. He gave one to his flatmate Mosher, who began poring over its contents. McIntyre says Mosher then called him. "I couldn't believe my ears. Mosh… asked me to confirm emails attributed to me – which I did. They didn't give me the email link." This version of events is consistent with Mosher's teasing claim, in a subsequent blog that "on the morning of Nov 19th two people held the file (that I know of). Me on a CD and a blog moderator who was holding the FOIA comment. Embargoed at the request of the blog owner… Did I download the files? No. How did you [I] get them? On a CD. Who gave them to you? Can't say." Mosher says at this stage he wasn't sure if the emails were genuine or an elaborate hoax.

On 19 November, McIntyre received an email from a regular correspondent to his blog site from the University of East Anglia. This was the head of the university isotope analysis unit, Paul Dennis. Dennis is a public advocate of greater data freedom, whose own researches on ice core data leave him unimpressed by more alarming speculation about climate change. He has been interviewed by police in connection with the alleged hack.

Dennis told McIntyre that CRU people were trying to secure their servers, following the discovery of a leak. This gave the bloggers the evidence they needed that the material they had was genuine. But Mosher checked again. "I called people mentioned in the mails. I read them mails. The actual person inside CRU had no clue what this message meant to me. He passed me no information, just told me what I needed to know." Whatever that was, it proved the emails were genuine. Courrielche adds: "Shortly after confirming the authenticity of the Climategate files, Mosher says he saw a link to the files on the Air Vent site. 'My first reaction was relief. I didn't want to be the only person who had those files.' " Nobody else – including Jeff id, off on his hunting trip – seems to have noticed. But having certified the veracity of the file, Mosher got to work.

He posted a comment to Liljegren's Blackboard blog, pointing to the Air Vent site. Liljegren then downloaded the files, and Mosher started posting them as emails onto her site. Within minutes, Gavin Schmidt, an occasional contributor to the Blackboard, was sending Liljegren emails warming that this could be illegal. But by now Mosher was posting emails onto McIntyre's ClimateAudit site, too. And half an hour later Watts, who was by now on his way back from Europe, gave Charles the Moderator permission to release the material onto his site. Since WUWT gets much more traffic than the others, this "broke" the story.

McIntyre categorically denies that he had any role in the copying of the files, or knowledge of who did. "Like many other readers of the various sites, I followed the pointers to Jeff id's site and downloaded the files on the afternoon of Nov 19. I was unprepared for what I encountered. Because I was intimately familiar with the context of so many of the emails, they were that much more shocking to me." After browsing, he says, he went off to play squash. He says: "To the best of my knowledge, neither Mosh nor CTM [Charles the Moderator] had (or has) the faintest idea of who assembled and released the dossier – other than speculations from their experience with computers. Nor do I. I talked to both Mosh and CTM on the late evening of 17th, when they were in the first throes of reading the emails. There is no doubt in my mind that they knew nothing of the source other than CTM knowing the Russian link."

Some remain suspicious of McIntyre. He is the only known participant in these early hours of the affair who was in the right time zone to have copied the files. And they point to a previous pattern that is strikingly similar to what happened in November.

On 24 July, as we saw in an earlier chapter, McIntyre received a big FOI refusal from CRU. The next day, he announced that he had got a mass of data from his mysterious mole. Again, in November, he received a big FOI refusal from CRU, and within days the "FOIA2009.zip" file was all over the web. There are other superficial similarities. McIntyre's July "mole" turned out to be an inadvertently public area of the CRU FTP server. The emails, according to the university, also came from that FTP server. But while McIntyre makes no secret of the fact that he was responsible for the first leak, albeit from a public place on the CRU server, he emphatically denies being behind the second. There is no reason to doubt his word. Moreover, many others could have been inspired to do in November what he did very publicly in July.

A number of other people claim to have stumbled on non-public files on the university server in the months before the hack. In December 2008 the British sceptic David Holland notified the university that "the search engine on your home page is broken and falling through to a directory". The university thanked him for letting them know and said it was caused by a "misconfiguration of the web server". Holland says he didn't download or alter anything since he knew it could be traced back to his computer.

Others were not so fastidious. In November 2009, Charles the Moderator blogged that "one day in late July I discovered they had left station data versions from 2003 and 1996 on their server – without web page links but accessible all the same. They were stale versions of the requested data… just sitting in cyberspace waiting for someone to download." After the July incident, CRU clearly tried to batten down the hatches. But perhaps they failed, either through technical failings or because someone inside was subverting the efforts.

So what actually happened in November? Charles the Moderator, along with Mosher, seems to have been closer to the perpetrator than anyone. Four days after the hack went public, he advanced his theory. "In the past I have worked at organisations where the computer network grew organically in a disorganised fashion. Security policies often fail as users take advantage of short-cuts… One of these is to share files using an FTP server… This can lead to unintentional sharing with the rest of the internet as noted in the [July] Phil Jones, CRU mole, example. Often the FTP server may also be the organisation's external web server. When this occurs, if the organisation does not lock down their network thoroughly, the security breaches which could happen by accident are far more likely to occur."

So, he argues, CRU inadvertently shared the emails file with others "by putting it in an FTP directory which was on the same CPU [central processing unit] as the external web server, or even worse, was on a shared driver somewhere to which the web server had permissions to access. In other words, if you knew where to look, it was publicly available. Then along comes our 'hackers' who happen to find it, download it, and the rest is history."

Charles the Moderator insists this is just a theory. But he is one of the few people who might be in a position to know if it is the truth of what happened. And if his theory is true, then the university will be left looking rather foolish. There will be no one to arrest.

CHAPTER 14
Tricks and lies

"How dare the world's media fall into the
trap set by contrarian propagandists?"

ELIZABETH MAY, LEADER OF CANADA'S GREEN PARTY, DECEMBER 2009

Within hours the news agenda of the hacked emails was set. Sceptics said the messages were a smoking gun that undermined the case that humans were causing climate change; that they revealed a conspiracy to suppress the truth about a gigantic hoax. And most of the media and political discussion since has been framed around that central proposition. The evidence for those claims came in the form of brief quotes publicised by climate sceptics, bloggers and lobbyists. In many cases, these have been taken out of context and twisted to mean something they were never intended to. Sometimes they have been given interpretations that are logically impossible. That didn't impinge much on the news.

Elizabeth May, veteran head of the Canadian Green Party, one of the first people to read all the emails, declared: "How dare the world's media fall into the trap set by contrarian propagandists without reading the whole set?" Quite so. If those journalists had read even a few words more, they would have realised that they were often being fed lies. Here, before we embark on an account of what the emails really reveal, are a few examples.

The most-quoted extract from the 1,000-plus emails is from Phil Jones to Mike Mann in 1999. In it, he discussed using "Mike's *Nature* trick" to "hide the decline". The phrase has been widely spun as an effort to prevent the truth getting out that global temperatures had stopped rising.

Politicians have been especially brazen. The Alaska governor Sarah Palin, in the *Washington Post* on 9 December, attacked the emailers as a "highly politicised scientific circle" who "manipulated data to 'hide the decline' in global temperatures". She was joined by the Republican senator and climate denier James Inhofe of Oklahoma. Inhofe attended the UN climate conference in Copenhagen in December 2009, as part of a Senate delegation. While there, he referred to Jones's "hide the decline" quote. He said: "Of course, he means hide the decline in temperatures." This is nonsense. Given the year the email was written, 1999, it cannot have been anything other than nonsense. At that time, there was no suggestion of a decline in temperatures. The previous year was the warmest year in the warmest decade on record. So what did Jones mean?

The full email from Jones says: "I've just completed Mike's *Nature* trick of adding in the real temps to each series for the last 20 years (ie from 1981 onwards) and from 1961 for Keith [Briffa]'s, to hide the decline." The decline being referred to was an apparent decline in temperatures shown in analysis of tree rings. Tree rings have previously correlated well with changes in temperature. But that relationship has broken down in the past half-century. The reasons are still debated. The "trick" was a graphic technique used by Mann in his famous 1998 "hockey stick" paper in *Nature*, to merge tree ring data from earlier times with thermometer data for recent decades. He explained what he had done in the paper. Jones was repeating it in another paper.

"This is a trick only in the sense of being a good way to deal with a vexing problem," Mann told the *Guardian*.

Clearly, this vexing problem with modern tree data raises genuine questions about older data – at least until the reason for the divergence is nailed down. But it is manifestly not clandestine data manipulation, nor, as claimed by Palin and Inhofe, is it a trick to hide global cooling. That charge is a lie.

But it is a lie that persists. Addicts of YouTube may have seen a natty animation of Mike Mann singing a song about "hide the decline", based on the song *Draggin' the Line* by Tommy James and the Shondells. No matter that the words "hide the decline" were Jones's and not Mann's. And only those quick to hit the pause button will have been able to check that the almost subliminal graphs flashed up on screen apparently showing a decline in temperatures almost entirely cover the period after the email was written.

While he was in Copenhagen at the climate conference, Inhofe linked the "trick" to "hide the decline" with the second most popular quotes from the emails. He said that "of course [Jones] meant hide the decline in temperatures, which caused another scientist, Kevin Trenberth of the National Centre for Atmospheric Research in Boulder, Colorado, to write: 'The fact is we can't account for the lack of warming at the moment and it is a travesty that we can't.'"

The link is bogus. The two emails were 10 years apart. Unlike Jones's, Trenberth's remark from October 2009 was indeed about the slackening of the warming trend since 1999 that some see as cooling. But Inhofe and others latched on to Trenberth's "travesty" phrase as a revelation showing that scientists were trying to keep cooling secret because it undermined their arguments about global warming. Again this is demonstrably false. Nothing was hidden – quite the contrary. For months, Trenberth had been discussing publicly his concerns about the inability of scientists to pin down the precise reason for the "absence of warming" since 1998. In particular, he wanted to know why 2008 was so cold.

He did not suggest that recent temperatures caused him to question the link between greenhouse gases and warming. But he had argued in the journal *Current Opinion in Environmental Stability* in early 2009 that "it is not a sufficient explanation to say that a cool year is due to natural variability". Such explanations "do not provide the physical mechanisms involved". This was the "travesty" he was referring to in his email. He wanted scientists to do better. Not to settle for a vague diagnosis of "natural variability".

Trenberth said the best way to improve the explanation and make it more specific was to make better measurements of the planet's energy budget. This would allow scientists to distinguish between any changes in the greenhouse effect, which would result in more or less heat overall in the atmosphere and oceans, and short-term natural cycles of variability, which merely redistribute heat. He was debating this with the former head of CRU Tom Wigley, who took a different view. But their genuine scientific discussion has, since the publication of the emails online, been hijacked by ignorant or malicious invective.

Other quotes from the emails were subjected to perverse or dishonest interpretations by commentators. Pat Michaels, the climatologist and polemicist for the rightwing Cato Institute, published a long op-ed piece in the *Washington Examiner*, slamming Mann for an email quote about keeping sceptics' papers out of the IPCC report "even if we have to redefine what the peer-reviewed literature is". Michaels is an old foe of Mann's, so it fed his agenda. But he should have read the email a bit more carefully. For this genuinely damaging statement was not made by Mann, as Michaels claimed, but by Jones. He was accusing the wrong guy.

In another case George Will, celebrated in some circles as an intellectual, told ABC's *This Week* programme in the US that Mann had said in an email that he wished to "delete, get rid of, the medieval warming period". No such words appear anywhere in the emails. What Mann said was that "it would be nice to try to 'contain' the putative 'MWP' ". And an intellectual like Will should have known that, in this context, "contain" means to understand its dimensions – how warm it was and how long it was. Mann explained as much to anyone who asked.

CHAPTER 15

A public disaster

*"We've never before seen a set of people accuse an
entire community of scientists of deliberate deception"*

SPENCER WEART, CLIMATE HISTORIAN, NOVEMBER 2009

The response of the science establishment to the emails scandal is
a case study in public relations disasters. One anonymous PR
figure from a major environment group said: "Their response will
be taught in university communications courses – because I'm
going to make sure it is."

When the scandal broke in the news media on Friday 20
November, the initial response from both the emailers and their
employers was to condemn the hackers and ignore what had been
hacked. In the face of a barrage of press inquiries, the University
of East Anglia put out a statement saying that "this information
has been obtained and published without our permission",
announced the police had been called in – and went home for the
weekend. The following week it added that "the selective publication of some stolen emails and other papers taken out of context
is mischievous and cannot be considered a genuine attempt to
engage with the issues in a responsible way". There were no interviews, no serious attempts to counter the gossip and malicious

innuendo spreading across the blogosphere about what the emails contained.

Phil Jones called most of the charges "ludicrous". A few, who knew what it was like, rushed to his defence. Ben Santer called Jones "one of the gentlemen of our field". He was standing firm in the face of "the vilest personal attacks" from "powerful forces of unreason". Kevin Trenberth said it was an attempt to undermine the Copenhagen climate conference. Mike Mann called the affair "a high-level orchestrated smear campaign to distract the public about the nature of the climate change problem". It was, he said, "a very professional job. There appears to have been a well-organised PR campaign that was ready to go at the time these emails were released. And that campaign involved all sorts of organisations that have lobbied against climate change legislation."

The science establishment agreed. In the US, the Union of Concerned Scientists called it a "manufactured controversy". Philip Campbell, editor-in-chief of *Nature*, told a Chinese journalist that so far as he was concerned the scientists had "behaved as researchers should" – an innocuous remark that came back to bite him. Bob Ward of the Grantham Institute for Climate Change at the London School of Economics agreed that Jones and Mann "have been subjected to a co-ordinated campaign of personal attacks on their reputation".

But whatever the motives were, the contents of the emails were not edifying. And the sceptics were making hay. They didn't need a "well-organised PR campaign" to get their message over. They gleefully sifted the emails and mined quotes that, superficially at least, revealed extensive data manipulation and scientific sleaze. The bloggers were at times obscene in their hatred and glee. A virtual lynch mob was at work – and not always so virtual. A picket formed outside the offices of NCAR in Boulder, where Tom Wigley, Trenberth and other emailers worked. Wigley was among a number of climate scientists who received death threats.

After Jones had revealed in an interview that he had contemplated suicide over the affair, one correspondent to the Free Republic website jeered: "Hear! Hear! I second that!.... oooooh... he's actually hurt people think so bad about him... ooooooh... they want him to die, to commit suicide... oooooohh... he's even thought about it... blah blah blah... blah blah blah blah... He *should* feel bad. He *should* quit. He *should* blow out whatever is left of the brains he once had. An evil, evil man who did everything in his power to enslave and cripple your children and mine."

Such stuff made even the more hysterical journalists appear moderate. The *Daily Telegraph* published a blog by a former news reporter, James Delingpole, claiming the affair "exposed the conspiracy behind the anthropogenic global warming myth", adding for good measure that "this scandal could well be the greatest in modern science". Such frothing could be dismissed.

But serious commentators realised something significant was happening. Four days after the leak, the environmental commentator George Monbiot stunned greens by writing in the *Guardian* that "there are some messages that require no spin to make them look bad... I believe Phil Jones should now resign." The mood changed. Even Mann, whose words featured prominently in early quotes from the emails, began an op-ed in the *Washington Post* with the words: "I cannot condone some things that colleagues of mine wrote." And later he told *Discover* magazine that "Jones... showed some poor judgment." RealClimate offered the half-apologetic insight that the emails offered "a peek into how scientists actually interact... we are sure it comes as no shock to know that many scientists do not hold Steve McIntyre in high regard." The post went on, "Gravity isn't a useful theory because Newton was a nice person."

But such guarded apologies didn't turn the tide. A survey in the US found that 49 per cent of respondents claimed to have followed news of Climategate "very closely or somewhat closely", and 59 per cent found it "very likely or somewhat likely" that some scientists hade falsified research data to support their own theories and beliefs about global warming. According to American science historian Spencer Weart, "We've never before seen a set of people accuse an entire community of scientists of deliberate deception and other professional malfeasance. Even the tobacco companies never tried to slander legitimate cancer researchers."

All sorts of people surfed the wave. In the US, sceptical physicists used the moment to revive a campaign to overturn a 2007 declaration by the American Physical Society that evidence of man-made climate change was "incontrovertible". Their letter began: "By now everyone has heard of... Climategate, which was and is an international scientific fraud, the worst any of us have seen."

In Britain, Lord Lawson launched the Global Warming Policy Foundation (GWPF) four days after the story broke. Lawson had been Chancellor of the Exchequer in 1989, when Margaret Thatcher assembled her entire cabinet to hear a seminar on climate change. Tom Wigley, then director of CRU, had been the star performer. But Lawson now opposes measures to fight climate change.

The foundation's stated purpose is to "bring reason, integrity and balance to a debate that has become seriously unbalanced, irrationally alarmist and all too often depressingly intolerant". But its trustees and academic advisers do not reflect that balance. Most have a public record making sceptical comments about climate science. Lawson has repeatedly refused to say who is funding his organisation. He told the House of Commons Climategate inquiry in March that some donors might want to preserve their anonymity because their partners or families might have different views.

Lawson appointed as the foundation's director Benny Peiser. He is a social anthropologist, part-time lecturer at the School of Sport and Exercise Sciences at Liverpool's John Moores University, and long-standing co-editor of *Energy and Environment*. The journal is trashed by Jones in the emails as "the worst journal in the world" for its patronage of what he regards as poor-quality papers by sceptics. Within hours of his appointment, Peiser had begun what became a frequent media presence on the increasing number of occasions, particularly on BBC News, when editors needed a sceptical voice in their climate coverage.

Meanwhile, the *Daily Express* in London splashed a long story across its front page headlined: "100 reasons why global warming is natural". It said the list came from a "dossier" issued by the European Foundation, a UK-based group that campaigns mostly against European integration. The author of the dossier, the foundation's journal editor, Jim McConalogue, told the *Express* that it "demonstrates how tenuous, improper and indeed false the scientific and political claims are for man-made global warming". To many, his report demonstrates those attributes among a certain class of politically motivated sceptics.

The "dossier" comprises fewer than 170 sentences of large type across 14 pages. The "reasons" that global warming is natural include questions, tendentious quotes from sceptics, non-sequiturs by the dozen, and the wonderfully circular no 9: "Since the cause of global warming is mostly natural, then there is in actual fact very little we can do about it." Most of the 100 reasons were either meaningless or scientific nonsense, according to *New Scientist* magazine, which gave up after debunking the first 50.

Despite making global headlines, Climategate had little impact at the UN climate negotiations in Copenhagen in early December. These were about raw politics and not climate science. Only one delegate mentioned the emails during formal proceedings. Saudi Arabia's lead climate negotiator, Mohammed Al-Sabban, claimed on the opening day that "it appears from the details of the scandal that there is no relationship whatsoever between human activities and climate change" – a view that not even the most sceptical scientists would endorse.

On the fringes, a few crowd-pleasers had things to say on the subject, including Danish environmental economist Bjørn Lomborg and Senator James Inhofe. Back home from Copenhagen, Inhofe wrote a report for his senate environment committee claiming that "scientists involved in the CRU controversy violated fundamental ethical principles governing taxpayer-funded research and in some cases may have violated federal laws". The report named 17 who turned up, however briefly, in the emails while also holding positions within the IPCC. Inhofe called for a criminal investigation into their activities.

Michael Oppenheimer of Princeton University, who appeared on the list, called it "a drag net... a McCarthyite tactic: pull in anyone who had anything to do with anyone because they happened to converse with some by email, and threaten them with criminal activity." Gavin Schmidt, also on the list, said: "The idea very clearly is to put a chilling effect on scientists speaking out in public and to tell others to keep their heads down." And some in the sceptical community would go further. "I serious believe we should kick then while they're down," says Marc Morano of ClimateDepot.com.

*

The story was not going away. So the University of East Anglia launched Plan B. Jones stood down as director of CRU, while the pro vice-chancellor Trevor Davies, who had himself been director of CRU from 1993 to 1998, launched an independent inquiry under a senior Scottish civil servant of breathtakingly dull probity, Sir Muir Russell.

But the policy of silence from the scientists persisted. When I phoned acting director of CRU Peter Liss, a marine scientist and an old contact, he told me sorrowfully that he could not talk and nor could his staff. They were not allowed to. On the eve of the *Guardian* publication of the first part of its investigation into the emails, I phoned Jones at his home. (The number was buried in the emails.) He told me he could not speak.

Jacquie Burgess, head of the university's School of Environmental Sciences, within which CRU nestled, told me that there was no outright ban on media interviews. But any of her staff approached by the media were "asked to seek guidance from the UEA communications team before doing so", and noted the university's "duty of care"to its staff. One UEA environment scientist, Andy Watson, broke ranks and agreed to be interviewed on BBC's *Newsnight*. "Despite the best efforts of the sceptics, there is no instance in these emails that anyone has found so far – and there are millions of people looking – that suggests the scientists manipulated their fundamental data," he said. (But sceptics watching reached for their blogs only when, thinking the microphone was dead, he called his on-screen adversary Morano an "arsehole".)

Otherwise, the silence was deafening – and disastrous. It simply left the field open for sceptics to dominate discussion of the emails. When the *Guardian* investigation ran in early February, the university at first made no response. Monbiot, in a blog a few days later, called for the departure of the university's head of communications, Annie Ogden, on the grounds that "she was warned repeatedly that the university's handling of this issue was a catastrophe and still the policy – of utter passivity in the face of crisis – remains unchanged".

Only after the *Guardian* articles were published did the university begin to encourage Jones to give occasional interviews. He spoke to *Nature*, the *Sunday Times* and, in a long email exchange, to the BBC environment analyst Roger Harrabin. A fight back, of sorts, had at last begun.

Jones admitted to *Nature* that his much-criticised failure to keep records about the locations of Chinese weather stations used in a major paper was "not acceptable". In effect, he conceded that Doug Keenan had been right in some of his criticisms of the notorious 20-year-old Jones/Wang paper. Jones said he might submit a correction to *Nature*. But he nonetheless attacked bloggers and other critics for "hijacking the peer review process… Why don't they do their own [temperature] reconstructions? If they want to criticise, they should write their own papers," he said.

The climate scientists' blog site RealClimate has taken up the defence of the emails and their authors with vigour, including critiquing the *Guardian*'s investigation. Those comments are quoted and examined extensively elsewhere in the book, so I won't repeat them here. Some bloggers were supportive, even keen to help with the rehabilitation. The pick was the "We Love You, Phil Jones" campaign run by the popular climate science blogger Jo Abbess. "Phil Jones, we are a small but growing group, who just want you to know that we love you, respect you, and we think you should not only keep your job, but be awarded an honour for bravery in the face of a huge and nasty smear campaign. Pay no attention to the *Guardian*…" it began.

And Elizabeth May, head of the Canadian Green Party, having read the emails, said that for her they "give a picture of thoroughly decent scientists increasingly finding themselves in a nightmare. They write each other in disbelief, protesting: 'I have never been political. I am an honest scientist.' "

But the response of most environmental groups was more muted even than that of the university. They kept their heads down. My anonymous environmental PR man said: "The emails represented a seminal moment in the climate debate of the last five years, and it was a moment that broke decisively against us. I think the CRU leak is nothing less than catastrophic." They simply did not want to be associated with it.

Greenpeace, Friends of the Earth, WWF – all carried on as if nothing had happened. Only one journalist that I know of commented on this strange silence. Britain's longest-standing environment correspondent, Geoffrey Lean, now of the *Daily Telegraph*, accused them of being "gutless" in the wake of the climate emails. "Their almost complete silence speaks volumes about their lack of courage... I have been told by a senior figure at Friends of the Earth that the group should only undertake campaigns where it already has public support." Green groups, he said, have "become part of the establishment, and let the sceptics take over their former role as insurgents. And they now cannot get their act together to respond." The green emperors had no clothes.

Perhaps Lean was alone in this observation because many science and environmental journalists did not pursue the story initially. Some were simply too busy preparing for the high politics of the Copenhagen climate conference, then only days away. Others agreed with Mark Lynas, a campaigner-cum-journalist, who argued in the *New Statesman* that reporting the emails was giving in to "barbarians at the gate", adding fuel to fires lit by a "denialist movement" orchestrated by "conservative thinks tanks and powerful industrial vested interests".

They certainly did not agree with Matt Ridley in the rival magazine the *Spectator*. The prolific science writer, climate sceptic and inheritor of his father's seat as chairman of the building society Northern Rock (until its spectacular collapse in 2007) argued that "while most of Fleet Street kowtowed to the green lobby", it was the blogging sceptics who "uncovered the spin and deception that finally cracked the consensus" in climate science.

Fiona Fox, director of the London-based Science Media Centre, came out in favour of the virtues of reporting uncomfortable truths. She criticised those correspondents who did not want to touch this grubby story in a blog in late January. "I think they are wrong. And I'm not sure they are doing science any favours." Many specialist journalists, she said, believe that climate change "is the most important issue of our times, and therefore requires a special kind of journalism – a kind of journalism that would actually argue against running stories on UEA-gate and Glaciergate; a kind of journalism that boasts that certain sceptics will never be interviewed on their watch; and a kind of journalism that asks whether stories might damage negotiations at a critical time."

But she said, "it is precisely because climate change is so important that it requires... the most rigorous, objective, robust, investigative journalism we can muster. Any reporting that gives mainstream climate science an easier ride simply plays into the sceptics' hands."

The problem was that two issues were getting confused: the reporting of the science and the reporting of the conduct of the scientists. The eager bloggers could not make the distinction, and nor could many editors. The BBC has rarely reported directly on the Climategate controversy, but it has nonetheless coloured the corporation's reporting of climate science. Viewers have often watched a crashing of editorial gears. For several years most of the BBC's coverage of climate change has been based on the scientific consensus that warming is real and that mankind is to blame. This had been reinforced by a study for the BBC Trust that concluded in 2007: "The weight of evidence no longer justifies equal space being given to the opponents of the consensus."

But even before Climategate, the BBC had been reviewing both its climate and science coverage. Deputy director-general Mark Byford organised an editorial seminar on climate change in September 2009. Insiders say the seminar followed an MP's complaint that sceptics didn't get a hearing and an in-house investigation that discovered that very often they did not. After the seminar, deputy editor of news Steve Mitchell sent round a memo advising of the "need to reflect deniers in run-up to Copenhagen".

One straw in the wind at the BBC was the awarding to a Leeds-based weather man with mildly sceptical views, Paul Hudson, of the title "climate correspondent" – to the chagrin of news correspondents covering that beat. In October 2009, Hudson began posting blogs on the BBC website about the absence of warming since 1998. This caused a flurry of concern among climate scientists, who began sending emails to each other discussing what looked like an outbreak of climate scepticism at the BBC. Some of their emails found their way back to Hudson. Hudson has since refused to divulge his source. So it remains unclear whether this was a trial run for the later leak, or simply a result of someone among the long list of people involved in the correspondence copying it to Hudson.

Regardless of events in Leeds, the ripples of the Mitchell memo spread through the corporation. Radio 4's Today programme raised eyebrows when, days before the Climategate emails leaked, it interviewed an Australian climate change denier, geologist Ian Plimer. He had what one critical insider called "the easiest of rides", with no cross-questioning for a string of highly contentious claims. And after Climategate the change became even more visible to viewers and listeners.

Reporters say they have been under pressure from editors to "get more sceptics on". One major beneficiary has been Benny Peiser from the Global Warming Policy Foundation, who has made repeated appearances on prime-time BBC news. "We are," one correspondent said privately, "back to the false-balance days that chiefs swore had been left behind".

The responses to the *Guardian* investigation of Climategate, published in early February, are a case study in themselves. I hope that we have fulfilled the classic reporters' function of investigating the truth. A part of that has been an attempt to break from the assumption of much media coverage that if any of the climate scientists are found to be at fault, then that undermines the edifice of science about climate change. It is not like that. As Gavin Schmidt put it: gravity isn't right because Newton was a nice guy.

We have sought to avoid tribalism, even over a cause that we do regard as among the most important of our time. Many other people who have no doubts about the scientific underpinnings of man-made climate change have been alarmed by the emails, and have applauded the *Guardian* for attempting to investigate them with rigour and impartiality, and to put them in their context.

One supporter in the US is Richard Littlemore in his Desmog-blog.com, which promises to "clear the PR pollution that clouds climate science" and which sceptics dismiss as "warmist". He praised the *Guardian* for tracking the story, for being "bluntly critical of the now-embarrassed climate scientists who featured [in] the emails". He said it "put the whole sideshow into context" by declaring that "nothing uncovered in the emails destroys the argument that humans are warming the planet".

Meanwhile a long analysis in the *Columbia Journalism Review*, the house journal of US journalism, praised the coverage as an example of what American newspapers should have done themselves. "It is this kind of detailed, intellectually honest reporting that news outlets like the *New York Times* should be striving for," it said.

Nonetheless, some climate scientists untouched by allegations in the emails or the *Guardian*'s coverage have been extremely angry at any criticism of their colleagues. Most notable has been Myles Allen, head of the climatic dynamics group at the University of Oxford. He wrote a comment for the *Guardian* accusing us of "hoping against hope to turn up a genuine error which fundamentally alters conclusions". And he sneered that we have ending up "looking rather desperate, complaining about procedures, picking holes in irrelevant software... The spectacle of journalists acting out their fantasies at the expense of both Phil Jones and their readers is looking increasingly pathetic." That's not how it feels from here.

Mark Lynas in the *New Statesman*, who spent two years poring over the work of climate scientists to produce his valuable book *Six Degrees*, likewise concluded that "even the liberal media establishment is falling over itself" to join the hounding of good climate scientists. He suggested that the *Guardian* did this "presumably to demonstrate its great rigour and impartiality... All paranoid conspiracists will be duly taken seriously, all climate deniers given their deserved moment in the sun." Sometimes the activities and views of such people have to be reported. We believe we have analysed their activities rigorously.

CHAPTER 16

Avalanche at the IPCC

"This seems to be a very important statement.
What is the confidence level/uncertainty?"

ANONYMOUS JAPANESE DELEGATE TO THE IPCC, 2006

Climate sceptics have been attacking what they see as the political remit of the IPCC for years. The institution responded by mending its working methods in small ways. After the "hand of man" row, and the assaults on Ben Santer in the mid-1990s, it sought to avoid ever again having to redraft a chapter after the writing of the summary. The hockey stick affair gave it some salutary lessons about the perils of promoting iconic graphs to the front of the document unless you were very, very sure about their probity. And for the fourth assessment, published in 2007, the decision was taken to archive internal discussions about drafts and to place the deliberations online for public use after the report itself was published.

But by the time of this most recent assessment, senior authors were not just mending their procedures, they were beginning to change the content of their reports, particularly the summary for policy-makers, to avoid disputes. During 2006, I discovered how lead authors of the IPCC assessment of the science of climate change – one of three reports by different working groups – were drawing in their horns to reduce the targets for sceptics. They were, I was told, taking great care not to make predictions that were not fully replicated in their own modelling studies.

This happened, insiders told me, partly because the chairmanship for the report had switched from Britain's Met Office Hadley Centre to the US's National Oceanic and Atmospheric Administration (NOAA) in Boulder, Colorado. There, atmospheric chemist Susan Solomon was in charge. She had won her scientific spurs in Antarctica in 1987 tracking down the chemistry of how man-made chlorine compounds were destroying the ozone layer. Now, back home in the American heartlands, she and her team had a reputation for being more sensitive than the Brits had been to attacks from climate sceptics. But their caution was also, no doubt, in part simply because the wider climate community had become better at seeing trouble coming.

Everyone I spoke to then connected with that report was looking over their shoulders at the sceptics. In mid-2006, as their drafts were going through the final stages, several authors told me they expected sceptics to attack any connection made in the IPCC report between global warming and increased damage from hurricanes. This of course was a year after Hurricane Katrina hit New Orleans. "I suspect the sceptics will want to try and dismantle the argument that there is a link," Mann told me. Santer agreed: "If I was an industry-funded sceptic, I'd hit that area hard, for sure." Trenberth, who had suffered a verbal battering for making the link the previous year, feared the worst. "I would not be surprised if the hurricane aspect of the report is targeted, along with my own role," he said.

I also spoke to sceptics. Pat Michaels told me that he would attack any attempt by the IPCC to link hurricanes to climate change. He said there was no modelling evidence for making such a link. Climate models could not reproduce small-scale and short-lived weather phenomena like hurricanes, however extreme. "[The IPCC] criticise people like me for saying the models are wrong, so it's going to be really interesting to see how they respond when their own people say the models are wrong," he said.

Some IPCC scientists whose day job was working for US government agencies, like NOAA and Trenberth's employer, the National Centre for Atmospheric Research (NCAR), feared the rows could escalate and threaten their federal funding. The US Senate's Environment and Public Works Committee, under its long-time chairman, the climate change denier James Inhofe, had begun investigating NCAR. Inhofe had repeatedly written to it, demanding details of financial and contractual arrangements with its scientists and with federal funding agencies such as the National Science Foundation.

In a letter to the NSF in February 2006, Inhofe said he needed the information to help him in "researching, analysing and under-standing the science of global climate change". That brought a wry laugh from the scientists. Mike Mann, who had been in Inhofe's sights for years, said: "Inhofe's actions appear to be an effort to discourage leading US scientists from being involved in international scientific assessment processes such as the IPCC."

Was this paranoia? It is hard to be sure. But a concerted attack from US legislators could have been disastrous for the IPCC. Out of 168 scientists listed as lead authors or reviewers assessing the science of climate change in 2006, 38 were from the US – more than twice as many as the second-largest national grouping, the British. Some fear Climategate may result in the attacks being renewed.

*

The IPCC report on the science of climate change was unveiled after a final plenary meeting in Paris at the start of 2007. Many observers saw signs of the new caution, particularly in what was missing. Some of the more scary scenarios discussed by scientists in recent years, involving positive feedback and "tipping points", were left on the cutting room floor. Others appeared in muted form in the body of the report, but failed to reach the crucial summary for policy-makers. In the past the summary's authors had been accused of hyping the science for effect, but this time round it looked as if they were deliberately downplaying things.

Journalists attending the launch conference searched in vain for mention of rising concerns about the potential for physical collapse of the Greenland ice sheet, rapid melting in Antarctica, a shut-down of the Gulf Stream in the Atlantic, or the release of carbon dioxide and methane from soil, the ocean bed and melting permafrost. All these potential disasters had been big topics in research papers. The British government had organised an international conference at the Hadley Centre in Exeter in the summer of 2005 specifically to discuss them. But for IPCC they were too hot to handle.

Richard Alley, a top US glaciologist who had briefed me in 2006 about potential Greenland ice collapses and had been working on the relevant IPCC chapter, said in Paris: "Our chapter will say that Greenland is doing things that could make it disintegrate much faster than people think… But we don't have a strong basis yet for projecting exactly what the ice sheets will do." So, he said, the summary excluded the new thinking.

Another IPCC author, the distinguished atmospheric chemist Venkatachalam Ramaswamy from NOAA, said that the panel's predictions of significant warming in far northern latitudes should give urgency to assessing potential methane releases from Siberia and the Arctic. But, he complained, his fears had failed to make it into the summary. "The chapters went through three sets of reviews," said Ramaswamy. "Anything qualitative rather than quantitative was knocked out. And if it got into the chapter, then the question was whether it would get into the summary. By and large where there was ambiguity or controversy, it didn't make it."

Most curious was the prognosis for sea level rise in a warmer world. The week before the report came out, Stefan Rahmstorf of Germany's Potsdam Institute for Climate Impact Research published a paper showing that world sea levels were now rising 50 per cent faster than predicted in the last IPCC report in 2001. He and co-author Jim Hansen suggested that this was the first sign of a dramatic acceleration of sea level rise likely in the coming decades, as polar ice sheets started to disintegrate.

Rahmstorf was an author for the IPCC chapter on sea level rise. But, despite knowing this paper was about to come out, the authors actually reduced their estimate of worst-case sea level rise in the coming century from 88cm to 59cm. As one scientist put it to me, real-world evidence was specifically excluded because it was not yet included in the models.

Solomon and her fellow authors evidently hoped this caution would take the heat out of disputes with sceptics. By concentrating on conclusions that they regarded as highly likely and downplaying less certain scenarios and more contentious ideas, they believed they had created a broad consensus. Achim Steiner, director of the UN Environment Programme, one of the IPCC's sponsors, was clearly on message when he sat beside Solomon at the launch and declared: "February 2nd will be remembered as the date when question marks were removed on whether climate change has anything to do with human activity – the moment when attention will shift to what on earth we are going to do about it."

For a while, it looked as if he was right. The main IPCC report on the science of climate change was attacked by the usual suspects, but no great rows developed. Then came Climategate. And then within weeks climate science suffered two more body blows. First there was the failure of the Copenhagen climate conference in December 2009. That did not damage the science as such but suggested that politicians might not be quite so scared by their scientists' warnings as they publicly suggested.

Then came revelations of a crass error in the latest IPCC assessment – not in Solomon's report on the science of climate change, but in a second parallel report on the likely impacts of climate change on society. Matters were made worse because the IPCC chairman, Rajendra Pachauri, at first rashly attempted to deny the error. While not directly linked to the emails, the saga of Glaciergate shows some similar failings in the science community – and a similar line-up of critics keen to make political hay.

Glaciergate was a bizarre affair. I found myself in a pivotal role, not just as a messenger of bad news but as an unwitting perpetrator. In December 2009, a Canadian glaciologist called Graham Cogley emailed me with a strange story. He said that the second of the IPCC assessment reports published more than two years before, on the likely impacts of climate change, contained a major error. It claimed that, thanks to global warming, the Himalayan glaciers would be gone by 2035. Cogley, a geographer from Trent University in Peterborough, Ontario, Canada, told me the 2035 date was extremely unlikely. "At current melting rates it would take at least 10 times longer," he said.

He added that the original source for the claim was one of my articles.

I was amazed. I remembered writing a story for *New Scientist* back in June 1999, quoting the comments of a leading Indian glaciologist called Syed Hasnain, vice-chancellor of Jawaharlal Nehru University in Delhi. In an interview, he had told me he had just written a report on Himalayan glaciology for a long-standing science body called the International Commission on Snow and Ice (ICSI). It was to be presented at the commission's upcoming meeting in Birmingham, England. His study was based on four years of research, and predicted that, at their current rate of decline, all the glaciers in the central and eastern Himalayas could disappear by 2035.

My published story, startling though it was, brought no reaction. I repeated Hasnain's prediction a couple of times in subsequent articles, and I noticed that Hasnain was reported making similar claims in other media. However, in 2005, I stopped using it, after British glaciologists tipped me off that it was unreliable.

In 2007, however, I noticed that the identical prediction appeared in the latest IPCC climate impacts report, though now covering the whole of the Himalayas and not just the central and eastern section. Presuming Hasnain's finding had now been validated to everyone's satisfaction, I used it again in a background article about climate science published prior to the Copenhagen climate conference in December 2009.

Cogley had spotted the piece. He assured me that there had in fact been no validation. And he had done some detective work. The only reference the IPCC report gave for the claim was a report written in 2005 by the Indian branch of the environment group WWF. That report included a quote that it attributed to Hasnain's ICSI report of 1999 and a reference to my article. Cogley told me he had spent some time getting hold of the Hasnain report, which had never been published, but only filed away by ICSI's then secretary.

When Cogley obtained it, he discovered that it did not contain the 2035 claim or anything like it.

It later emerged that the WWF's supposed quote from the Hasnain report had in fact been lifted from a 1997 article in the Indian environment magazine *Down to Earth*, which had been written with the help of Hasnain and published two months before mine. To all intents and appearances, the IPCC authors had simply cut and pasted that quote a second time – into their own report. In full the IPCC statement said: "Glaciers in the Himalayas are receding faster than in any other part of the world and, if the present rate continues, the likelihood of them disappearing by the year 2035 and perhaps sooner is very high."

This was error piled on top of error. A paragraph from a magazine article had been twice cut and pasted before turning up unchanged in an IPCC report, with a reference trail to a different magazine article, mine. Clearly nobody at the IPCC had checked anything. They couldn't even get the sourcing right.

This was a major error for an organisation that its former chairman Bob Watson describes as "one of the most rigorous scientific review bodies in existence". And it wasn't a small mistake. Cogley told me all the available evidence suggested that for the Himalayan glaciers to melt away would take at least 300 years. He wasn't arguing that climate change was any the less. Melting was happening fast all along the world's highest mountain range. But glaciers up to half a kilometre thick were not going to disappear overnight. He wondered if someone somewhere had mistyped 2350, a date that appeared in another report in the mid-1990s, as 2035. But that was speculation.

There was something else even more embarrassing for the IPCC. In November 2009, the Indian government had published a study of the state of Himalayan glaciers by Vijay Raina, another leading Indian glaciologist. He also dismissed any possibility of an imminent death for Himalayan glaciers. India's environment minister, Jairam Ramesh, who wrote the preface to the report, used it to accuse the IPCC authors of being "alarmist".

The IPCC's chairman hit back. Pachauri publicly denounced the Indian report as "voodoo science" and pointedly said that it was not based on peer-reviewed evidence. "We have a very clear idea of what is happening" in the Himalayas, he said. He clearly hadn't checked his own sources before making this charge against the minister. He had no idea it was his own report that lacked proper sourcing and peer review. If anyone was purveying voodoo science about Himalayan glaciers, it was Pachauri. As a result of his rashness, what began as a reprehensible lapse in report writing had become a diplomatic incident. But at that moment, only in India.

All this I had in front of me when I wrote a new article for *New Scientist* in January 2010. It was mildly embarrassing to realise an article of mine from a decade before had been the cause of such trouble. But I had simply been reporting what a leading glaciologist had said was in his upcoming report to a respected international science body. Hasnain's behaviour had been curious, to say the least. But the root of the problem clearly lay in the lamentable failure of IPCC authors to conduct even the most basic checks of what they were publishing as a scientific "assessment".

In any case, a decade after my original story, I had a clear duty to reveal what had happened. So I did. Before writing, I emailed the UN author behind the chapter concerned. He was yet another Indian glaciologist, called Murari Lal, the chairman of the non-profit Climate, Energy and Sustainable Development Analysis Centre in Ghaziabad, India. He told me he "outright rejected" the notion that the IPCC was off the mark on Himalayan glaciers or had not properly checked the prediction. "The IPCC authors did exactly what was expected from them," he said. Backing off slightly, he said that "the error, if any, lies with Dr Hasnain's assertion and not with the IPCC authors."

I emailed Hasnain, who, it turned out, had retired from his university post and now worked on studies of glaciers for Pachauri at the Energy Resources Institute. Hasnain insisted that the 2035 claim had been merely his speculation. That was news to me. And he blamed the IPCC for misusing a remark he made to a journalist. "The magic number of 2035 has not [been] mentioned in any research papers written by me, as no peer-reviewed journal will accept speculative figures... It is not proper for IPCC to include references from popular magazines or newspapers." I included both quotes in my story.

Once the new story appeared in *New Scientist*, it was picked up by other news media, particularly the *Sunday Times*, and was around the world in days. I subsequently discovered that several scientists involved in the IPCC review process had questioned the 2035 date before it was published in the 2007 report. Among them was Georg Kaser, a glaciologist from the University of Innsbruck, Austria, and a lead author of a chapter covering glaciers in the IPCC science assessment report. "I scanned the almost final draft at the end of 2006 and came across the 2035 reference," he told me. He queried the reference, he said, but it appears to have been too late in the day for it to be changed.

Online IPCC archives show that during the formal review, a delegate from the Japanese government also questioned the 2035 claim. He commented: "This seems to be a very important statement. What is the confidence level/certainty?" He didn't get an answer to his question. It was only at this stage that a reference to the WWF report was added, presumably by way of response. But the statement itself went unchanged into the final report and lay apparently unchallenged for more than two years until Cogley started asking questions.

Within days of my story breaking in January 2010, Pachauri conceded that "the clear and well-established standards of evidence required by the IPCC procedures were not applied properly". But he said it was a one-off, a single unfortunate paragraph in 10,000. But that was never going to be the end of the matter, especially as Pachauri himself had defended the paragraph to the Indian government.

Many journalists were by now hunting the IPCC. After the Climategate email revelations, it was open season on climate science in general. Some indulged in crude abuse. Fox News host Glenn Beck told his viewers: "If the IPCC had been done by Japanese scientists, there's not enough knives on planet Earth for the hari-kiri that should have occurred. These guys have so dishonoured themselves." Even those with a more forensic approach sensed that the IPCC report on climate impacts was a soft target. It seemed to rely more than the others on "grey literature" that had not been subjected to scientific peer review.

A series of revelations ensued. Most were relatively small beer. The co-chair of the impacts report, Martin Parry of Imperial College London, took time during a retreat on the Caribbean island of Montserrat, where he was writing a book, to email his fellow authors with the reassurance that it was "a clamour without substance".

Some of the issues raised fitted that description. For instance, the report noted in passing that 55 per cent of the Netherlands "is below sea level". Not true. The real figure is 26 per cent, though 55 per cent is at risk of flooding. The wrong data was provided by the Dutch government. The mistake had no bearing on other statements.

But others suggested sloppiness at the least. A chapter on Latin America predicted that "up to 40 per cent of the Amazon forests could react drastically to even a slight reduction in precipitation". This, like the glacier claim, was sourced to a report by WWF that had not been peer-reviewed. That was fair enough, except that the WWF report cited a paper by leading Amazon researcher Daniel Nepstad of the Woods Hole Research Centre in Falmouth, Massachusetts, that did not substantiate the claim.

It looked bad. Another Glaciergate, perhaps. But there the parallel ended. Later work by Nepstad did justify the 40 per cent prediction, as did modelling work by the UK Met Office and others. Nepstad said "the IPCC statement on the Amazon was correct." It turned out to be a case of lazy sourcing rather than flawed science.

Arguably more damaging was a statement in the report linking the growing economic impacts of natural disasters to climate change. The report's supplementary material, published online and in the report CD, contained a graph showing rising levels of damage from natural disasters corresponding to rising temperatures since 1970. The caption said this was to be published in a forthcoming paper by British risk consultant Robert Muir-Wood. But when the paper appeared the following year, it did not include the graph. Muir-Wood told me: "The graph could be misinterpreted and should not have been included [in the IPCC report]." It could simply reflect the fact that more people and more property were in harm's way, rather than being a result of any increase in the intensity or frequency of natural disasters.

And inevitably there was a fuss about hurricanes. The IPCC assessment of the climate science had been careful to avoid directly linking climate change and hurricanes. And the report on climate impacts largely followed that formula. Except once.

At one point, the report noted that research showed increasing damage from hurricanes in the Caribbean, but that this was due to changes in demography and wealth rather than climate change. The report correctly sourced that conclusion to a paper by Roger Pielke Jr of the University of Colorado, a noted climate sceptic. So far, so cautious. But then the chapter's authors seemed to chance their arm. They added a clause saying: "But this was before the remarkable hurricane losses of 2004 and 2005." It appeared to be almost a throwaway line, but it questioned Pielke's findings.

During the review process, one expert had asked for this clause to be justified or removed. But the chapter authors left the sentence intact. One responded online to the request by saying: "I believe Pielke agrees that adding 2004 and 2005 has the potential to change his earlier conclusions." Pielke has recent blogged that he was not asked and believes no such thing.

Several problems in the report relate to the chapters on particular continents, where it seems fewer scientists were involved and less peer-reviewed literature available. The glacier mistake happened in the Asia chapter; the Amazon sourcing mistake was in the Latin American chapter. And I uncovered a couple more issues in the Africa chapter.

The summary to the Africa chapter looked at the future of farming on the continent and highlighted a finding that "projected reductions in [crop] yield in some countries could be as much as 50 per cent by 2020". This captured public attention and appeared in speeches by Pachauri. But was it a fair reflection of the science on the topic?

The source for the conclusion was given as a report that had not been peer-reviewed, though it was written by a known expert, Moroccan Ali Agoumi, and funded by the US Agency for International Development. Citing the paper was legitimate. But there is doubt about how much can be concluded from Agoumi's study. The 11-page report, *Vulnerability of North African Countries to Climate Changes*, covered only three countries: Morocco, Tunisia and Algeria. It simply asserted at one point, without identifying specific evidence or a peer-reviewed source, that "studies on the future of vital agriculture in the region have shown the following risks, which are linked to climate change". It went on to list one risk as "deficient yields from rain-based agriculture of up to 50 per cent during the 2000-2020 period."

This was thin evidence, and the IPCC in places treated it carelessly. The authors did not acknowledge that Agoumi said the projected lost yields were "linked" to climate change, rather than necessarily caused by it. And in the executive summary of the Africa chapter, they even left out the fact that Agoumi's prediction applies only to rain-fed agriculture. This is an important distinction. In arid North Africa much crop production is irrigated. Bob Watson, former IPCC chairman, wrote in February 2010 that the synthesis report "failed to include the nuances" of the matter.

Drought is a major fear for farming in Africa, and elsewhere the Africa chapter looked in detail at water supplies. It said that by the 2020s "between 75 and 250 million people [in Africa] are projected to be exposed to an increase in water stress due to climate change". That is, those people will have less water to grow their crops and run their households. This finding was given headline status in the IPCC presentation of the report when it was launched in Brussels in early 2007, where an editing error meant they said this would happen by 2020 rather than 2025, the date in the paper it cited.

That paper was a report by Nigel Arnell, director of the Walker Institute for Climate System Research at Reading University, published in the journal *Global Environmental Change* in 2004. Arnell had run 14 climate scenarios on six models to predict how many people would experience "an increase in water stress" in 21 regions of the world. The Africa regions' totals for each scenario ranged from 74 to 239 million people. Hence, with rounding, the IPCC prediction.

But the IPCC interpretation of Arnell's paper was nonetheless questionable. The report ignored another table, given equal prominence in Arnell's paper, showing the numbers of people expected to experience "a decrease in water stress" – in other words those who will have more water than before, thanks to climate change. Subtracting that figure from the figure for those suffering increased water stress would reduce the apparent threat by almost three-quarters. That would clearly be wrong, however. As Arnell says, "an increase in water availability in one region does not offset a decrease in another".

But both Arnell and IPCC chapter co-author Coleen Vogel, of the University of Witwatersrand in Johannesburg, agreed in interviews that such a calculation would be important in answering the question insistently raised by the chapter: "Can African feed itself?" So to leave out the other figure altogether seems misleading.

That doesn't mean Africa faces anything other than a testing time over water. But to some it was another sign of sloppiness and an apparent tendency to exaggerate and cherry-pick data. Parry adamantly denies this. "There is a huge amount that is not known about the potential impacts from climate change," he says. "We have always tried to be clear about uncertainties in our current knowledge."

There has been a lot of soul-searching at the IPCC over all this. Some scientists have seen the criticism as politically inspired and best ignored, or have answered it defiantly. But others want to put their house in order. When the Glaciergate story broke, Parry's successor as chairman of the next impacts report, Chris Field of the Carnegie Institution for Science in Washington, contacted me to ask for details about the affair, so he could avoid it happening again. After asking questions of other scientists involved in post-mortems of the report, I have found myself copied into detailed discussions about possible reforms to the review system. This felt to me refreshingly different from the protestations of many scientists involved in the emails saga that they had done nothing wrong and needed to change nothing about the way they worked.

Obviously the IPCC will make mistakes. But as the gold standard of climate science, its reputation is vulnerable to any errors. What has damaged it most, however, has been its inability swiftly to rectify those mistakes. A big question in the discussions post-Glaciergate has been how to make corrections more quickly. The current problem is that, since reports are signed off in detail by committees of experts and government, any corrections should in theory be dealt with in the same way. Some believe it should establish a "rapid response" committee to handle such things.

That could be one outcome of a review into IPCC methods set up in March 2010. The six-month review will be conducted by the InterAcademy Council, a coalition of 15 national academies of science from round the world – including Britain's Royal Society. Launching the inquiry, Pachauri said: "We have received some criticism. We are receptive and sensitive to that and we are doing something about it."

Some believe that, 22 years after it was set up in the wake of the Villach meeting, the IPCC's heroic days are behind it. The panel has won a Nobel Prize and can be proud of its achievements. Maybe it should retire gracefully. More likely it will battle on. But in the wake of the leaked emails and Glaciergate, its reputation, like that of climate science, has taken a battering from which it may never fully recover.

CHAPTER 17
The inquiries

"It hasn't been standard practice to do that.
Maybe it should be, but it's not"

PHIL JONES ON SHARING DATA, MARCH 2010

By mid-February 2010, the Climategate affair was in danger of drowning in inquiries. There were three academic reviews under way, two commissioned by the University of East Anglia. In addition, a committee of MPs had begun its own hearings, which would include an assessment of whether the university had got the terms of reference for its inquiries right. And both the police and the Information Commissioner's office were investigating possible criminal behaviour – in the original "theft" of the emails and in what the emails reveal about how scientists and the University of East Anglia handled freedom of information requests.

Things were not going smoothly. The six-man panel of independent experts, set up by UEA to look for scientific wrongdoing at CRU, gave a press conference at the Royal Institution in London. But within hours of their first public appearance, one of the members had resigned after the Bishop Hill blog revealed he had given an interview to Chinese radio that appeared to pre-empt the inquiry by granting Jones and company a clear bill of scientific health. Philip Campbell, editor-in-chief of *Nature*, which had frequently published the work of many of the emailers and used them as reviewers, said: "I made the remarks in good faith, on the basis of media reports of the leaks. As I have made clear subsequently, I support the need for a full review of the facts behind the leaked emails."

Steve McIntyre and other sceptical bloggers tried to get a second member, Geoffrey Boulton, removed from the panel for allegedly having partisan views about climate science and a long association with the university between 1968 and 1980. They failed.

The panel was headed by former Scottish civil servant Sir Muir Russell, most famous north of the border for being in charge of the massively over-budget construction of the new Scottish parliament building at Holyrood. His inquiry, he said, was restricted to investigating whether CRU scientists were guilty of "manipulation or suppression of data", whether they complied with freedom of information legislation, and whether the unit's policies for "acquiring, assembling, subjecting to peer review and disseminating data and research findings" met best scientific practice.

Sir Muir insisted that, though the panel's work would be funded by the university, it was wholly independent. It would also not look at the science. "There are some very straightforward integrity issues to resolve," he said.

Many doubted whether this position could hold. After all, Jones said he rejected papers sent to him for peer review because the science was bad. How do you judge whether he was right without looking at the science? Likewise, UEA said it rejected requests for data because they were "vexatious", meaning they had no scientific purpose. And that is a defence in law. But again, how can this defence be judged without assessing the science?

The university chose the same day as the Russell press conference to announce that it had instituted a second inquiry to conduct a "reappraisal of the science in the Climatic Research Unit's key publications". In late March it emerged that the inquiry would be headed by eminent geologist Lord Oxburgh with a team of six international scientists covering physics, meteorology, statistics, technology, earth sciences and the environment.

Oxburgh was a former chairman of both the House of Lords science select committee and the oil giant Shell. Six years before, on taking up his Shell post, he told the *Guardian* that he saw "very little hope for the world" without the development of technology to capture carbon emissions. As the *Financial Times* pointed out, he also "has strong business interests in biofuels, is chairman of the wind company Falck Renewables, and a board member of Climate Change Capital, a major investor in carbon credits". Would sceptics try to argue that such credentials should rule him out of investigating the science? Well, yes, they would.

It also emerged in February that investigations were set to continue at Penn State University in the US into possible research misconduct by Mike Mann. A preliminary inquiry, involving reading 206 emails written by Mann and a further 171 either sent to him or mentioning him, found that Mann had no case to answer on four charges of falsifying or suppressing data, deleting emails and misusing confidential information. But it said there should be a formal investigation into whether the emails revealed Mann was guilty of "seriously deviating from accepted practice within the academic community". Mann called the findings so far "a vindication".

The university's inquiries were sedate and secretive. The MPs were in a hurry and happier to attract headlines. With a general election just months away, they had only till early April to complete their work. Their request for evidence brought in more than 50 submissions. They ranged from the irrelevant (a woman aggrieved about getting bad exam results at the University of East Anglia, who thought this exposed wider failings on campus) to the commercial (the world's largest coal mining company, Peabody Energy, offered its thoughts), and from the forensic (mostly unpicking Jones's work) to the testimonial ("Keith Briffa is a man who encourages vigorous debate rather than suppressing it," said Exeter climatologist and IPCC author Peter Cox.)

The most eye-catching evidence came from the Institute of Physics, a London-based body of 36,000 physicists from round the world. In 13 coruscating paragraphs, its director of science, Peter Main, wrote of "*prima facie* evidence of determined and coordinated refusals to comply with honorable scientific traditions and freedom of information law", "manipulation of the publication and peer review system", and "intolerance to challenge… which is vital to the integrity of the scientific process".

The internet was soon humming with claims that the institute's submission had been hijacked by energy industry lobbyists, and had never been properly discussed by its science board. The *Guardian* front-paged a story that its evidence had been written with help from "an energy industry consultant who argues that global warming is a religion". The man in question, Peter Gill, a member of the institute's energy management group, heads an energy consultancy based in Dorking called Crestport Services. While it appears the institute did email members of its science board with the draft statement, only three responded and it is not clear how many read it. The institute denied that it was "part of some anti-climate change agenda. We believe the basic evidence on climate change is clear and well established, and the case for taking action now to mitigate its effects is undeniable."

Other scientific bodies also raised concerns about how CRU did its science. The Royal Society of Chemistry noted that the "apparent resistance of researchers from [CRU] to disclose research data… may infer that the scientific results or methods used are not robust enough to face scrutiny. This has far-reaching consequences for the reputation of science as a whole."

The Royal Statistical Society said, "It is well understood that peer review cannot guarantee that what is published is correct. The best guarantor of scientific quality is that others are able to examine in detail the arguments that have been used [and] that experiments and calculations can be repeated to verify their conclusions. If data, or the methods used, are withheld, it is impossible to do this… The crucial step in improving the quality of the debate on global warming will be to place the data, the analysis methods and the models in the public domain."

The MPs' oral hearings, held on 1 March, were mostly a disappointment. The questioning was occasionally challenging, but limply unforensic. The exchanges did produce some insights and some theatre. The star turn, of course, was Jones, whose performance I reviewed at the start of this book. He was the anti-hero. But the true pantomime villain, sitting beside him throughout his interrogation, was his vice-chancellor, Edward Acton, whose repeated interruptions and near-permanent smile, while clearly intended to reassure the grim-faced Jones, evidently irritated the MPs.

When Acton asked to begin with a preamble, he was simply told that no, he couldn't. When he butted in to the questioning of Jones to tell the MPs helpful things like "I think that answer was spot on", he was simply ignored. When he was allowed to speak, he could be economical with the truth. For instance, he defended

CRU's failure to deal promptly with freedom of information requests during 2009 by saying it had only three "full-time" academics. But its website lists a director and deputy director, a research manager, three "academic staff" and six research staff, as well as longer lists of visiting fellows, associate fellows, support staff and postgraduate students. Many of them, like Jones, Briffa and Osborn, have been there for more than a decade.

What he might have meant was that only three of CRU's staff are paid for by the university as environment faculty members. The rest are employed according to the unit's long-term contracts. These include its work on global temperature data, which has been largely funded for the past two decades by the US government's Department of Energy.

Acton was smooth and slippery. He justified his suggestion that the government Information Commissioner had cleared university staff of acting illegally over freedom of information requests by saying the commissioner had only reported "*prima facie* evidence [which] isn't the same as you've been found to breach the law".

MP Graham Stringer, himself a chemist, was more sceptical and acerbic than his colleagues. He asked whether Acton's submission to the inquiry, exonerating Jones from any misdemeanours, might not prejudge his own Russell inquiry. Acton replied that he had no evidence that the science was flawed. But since the Muir Russell inquiry is not tasked with looking at the science, but only at the scientific process, that amounted to answering a question that had not been asked, and ignoring the one that had been asked.

As the questioning proceeded, Jones at times became candid almost without meaning to. Why, he was asked, did he tell the Australian freelance researcher Warwick Hughes that he wouldn't hand over data because "you only want to find something wrong with it"? Jones replied: "We had a lot of work and resources tied up in it." His nervousness encouraged him to answer questions that had not quite been asked. "I've written some pretty awful emails," he admitted at one point. It was the only time the smile left Acton's face. And the journalists had their headline.

Jones insisted that what he did, for good or ill, was what his fellow climate scientists did. They didn't publish all their data and methods because "it hasn't been standard practice to do that. Maybe it should be, but it's not." Following that came the most startling observation, when Jones was asked how often scientists reviewing his papers for probity before publication had requested to see details of his raw data, methodology and computer codes. "They've never asked," he said. The rigour of peer review came crashing down before our eyes.

Since they were holding just one session of oral hearings, the MPs raced through their agenda. Even given the shortage of time, their choice of interviewees was odd, showing a strong bias towards the "great and good" of parliament and government, rather than the real adversaries. Lord Lawson, a latter-day climate sceptic, was first up. In his haste to get his story over, he barely bothered with the questions from what *Guardian* sketch writer Simon Hoggart called "pipsqueak MPs".

Lawson called the hockey stick "largely fraudulent", the University of East Anglia's submission to the MPs "appalling" (for describing peer review rather than full disclosure as the "keystone" of scientific self-policing), and accused the MPs of deviating from their terms of reference in daring to ask how his thinktank is funded.

Anticipating Jones's later insistence that he always did his best to answer freedom of information requests, Lawson noted scornfully that "proper scientists, scientists with integrity, wish to reveal their data and all their methods. They do not require freedom of information requests." How did he know? Well, he dealt with scientists quite a lot when he was secretary of state for energy, he said. That was 27 years ago. But the MPs were in awe nonetheless. None of them had been secretary of state for anything.

Sitting beside Lawson was Benny Peiser, the head of the noble lord's recently created thinktank, the Global Warming Policy Foundation. Peiser barely got a word in apart from the pious-sounding observation that "personally I don't think the disclosures make a big change to the overall scientific debate, but people are disillusioned... It's tarnished the image of British science round the world."

Then came Richard Thomas, until June 2009 the government's Information Commissioner and yet to be replaced. He was unhelpful, sticking to generalities and unable to answer any questions about events since his departure, which came a few weeks before FoI requests to Jones and his colleagues hotted up. "I don't have the information," the information officer said more than once. Thomas was followed by Muir Russell, who dead-batted questions about his own inquiry. Finally came the big beasts from the labs: the chief scientists for the Met Office, the environment department and the government as a whole, respectively Julia Slingo, Bob Watson (a former chairman of the IPCC) and John Beddington.

Beddington had initially responded to the emails by warning scientists that "clarity and transparency must be reinforced to build and maintain trust – internal and external – in science." But he did not want to enlarge on that. Instead, he backtracked, arguing that transparency had to be achieved without "demotivating" scientists. All in all, he did not get much beyond the splendidly pompous first paragraph of his written submission: "Whilst no system can be perfect, I believe the integrity of British science stands of the highest order... It would not be appropriate to comment on the specific circumstances of the University of East Anglia... before Sir Muir Russell has reported."

It was a dull end. No sign of Canadian Rottweiler mathematician Steve McIntyre, the arch-villain of dozens of the CRU crew's emails. Or of Doug Keenan, who had accused Jones of fraud in a peer-reviewed journal. Without Jones's most tenacious adversaries even in the room, let alone answering questions, the plot fizzled out at the end of the first act.

The MPs were in a rush to finish their report ahead of the election. When they published their findings on the last day of March, they admitted that they did not go into many of the issues they would have wished. The report says: "We would have preferred to carry out a wider inquiry into the science of global warming itself." Chairman Phil Willis told the *Guardian* they had put many of the more forensic questions to one side. And McIntyre and his fellow inquisitors went unmentioned throughout.

The MPs gave Jones an easy ride. They were of course right to absolve Jones and his colleagues of many of the charges that bloggers have thrown at him. But, in their rush to judgment, they avoided examining more forensic charges of malpractice, including those raised by the *Guardian*. At their final drafting meeting, only Graham Stringer voted against the sentence: "Within our limited inquiry and the evidence we took, the scientific reputation of Professor Jones and CRU remains intact."

But it was a limited absolution. They agreed that it must have been "frustrating" for Jones to handle freedom of information requests from people "he knew – or perceived – were motivated by a desire simply to undermine his work". But they said that his "blunt refusals to share data, even unrestricted data" led to "unfortunate email exchanges" and was "inevitably counterproductive".

The MPs sometimes got confused. They accepted Jones's claim that CRU's addiction to secrecy was "standard practice" among climate scientists. At the launch press conference, the committee chairman Phil Willis called this custom "reprehensible". Yet they also noted that NASA scientists doing similar work were much more open. And they never asked whether, as one of climate science's senior figures, Jones should take some responsibility for what standard practice was. Nor did they hold him to account for the "unacceptable" culture within CRU, where he has worked for more than 20 years, been its most high-profile researcher for more than a decade, and its director for six years.

The MPs did not accept the claim from CRU and the university that their staff had failed to deal properly with freedom of information requests because they were simply overwhelmed by the number of requests. Rather, they found that CRU's "unhelpful approach" to legitimate requests led to them "multiplying", while the university did not offer the right administrative support.

Perhaps more at home with administrators than scientists, the MPs felt on stronger ground chastising the university administration for presiding, and sometimes encouraging, a "culture of withholding information" within CRU, a culture they found "unacceptable". The university had been guilty of a "failure to grasp fully the potential damage [from] non-disclosure of FOIA requests". Its information officers colluded with CRU to subvert legitimate freedom of information requests, and "found ways to support" the culture of secrecy. In a key statement that not even the proliferation of acronyms can disguise, they say: "We must put on record our concern about the manner in which UEA allowed CRU to handle FOIA requests." Quite so, given the *prima facie* case for its occasional illegality. "We recommend that all publicly funded research groups consider whether they are being as open as they can be, and ought to be, with the details of their method-ologies," the MPs said.

All this, incidentally, put the official university inquiry under Sir Muir Russell in an odd position. The university has asked it to investigate possible misdemeanours at CRU. But the MPs were suggesting that it was the university itself that was at least as much to blame for the Climategate debacle.

The MPs gave Sir Muir a great deal of food for thought all round. They agreed with Lord Lawson that his inquiry should conduct its interviews and hearings "in public wherever possible". Unless Sir Muir had spoken to nobody in the previous four months, he evidently was not doing that. They said his inquiry should "publish all written evidence on its website as soon as possi-ble". As the MPs' report appeared, it had been a month since the deadline for submissions to his inquiry had closed, yet none had been posted. And they suggested that a test of how truly inde-pendent the inquiry was would be whether it gave the UEA an advance copy, which was the existing plan.

Worse, the MPs gave Sir Muir a long list of things to investi-gate or rule on, things the MPs had not had time to pursue. They included most of the hard topics, including deciding whether emails were deleted in breach of freedom of information law; coming up with rules for CRU on sharing data; and deciding whether Jones "subverted the peer review process". As they headed back to their constituencies to face re-election or retire-ment, the MPs had safely washed their hands of the affair.

While Sir Muir lumbered on, Lord Oxburgh's inquiry into the probity of the science conducted by CRU held a couple of meetings and published its final eight-page report in mid-April, within a month of being formed. Its headline finding was that there was "no evidence of deliberate scientific malpractice". But it had some significant barbs for this "small group of dedicated if slightly disorganised researchers". CRU's work in both tree ring analysis and collating thermometer data depended on elaborate number-crunching, so "we cannot help remarking that research in an area that depends so heavily on statistical methods has not been carried out in close collaboration with professional statisticians", it noted archly.

As mentioned earlier, it found poor documentation over the years of the host of inevitable decisions on the use of both thermometer and tree ring data that depended on judgment calls and methods that were "somewhat subjective". People who questioned the integrity with which those decisions were taken were "uncharitable", Oxburgh said. "The CRU sins were of omissions rather than commission." But it gave a firm nod of support to the more civilised of CRU's critics by pointing out that their questioning "will ultimately have a beneficial effect and improve working practices". It sounded like a lofty call for everyone to shape up – and grow up.

CHAPTER 18

Bonfire of the scientific vanities

"Einstein did not start his career at Princeton,
but rather at a patent office"

JUDY CURRY ON THE NEED TO OPEN UP SCIENCE TO SCEPTICS, NOVEMBER 2009

The process of scientific discovery requires producing findings that others can test by trying to replicate or falsify them. That is how good theories are bolstered, how bad theories are discarded, and how knowledge is advanced. Some research is easy to replicate. All you need is a lab. Some is more complicated. And not much is more complicated than research that relies on huge amounts of data assembled from all over the world over many decades. Both Jones's temperature data for the past 160 years and Mann's proxy data of temperatures before that fall into this category.

The difficulty in replicating this work is one reason that sceptics, rightly or wrongly, have been able to claim that bad climate science has proliferated. That is why the demands have grown for scientists to release their raw data, even to people outside the research community. But the complexity of the work is also why researchers who have spent years, sometimes decades, assembling and analysing their data are unwilling to hand it out to the first blogger to ask for it under a freedom of information request. It is "their" data, analysed using "their" protocols and computer codes. As Jones notoriously put it to Warwick Hughes: "Why should I make the data available to you, when your aim is to try and find something wrong with it?"

As Myles Allen of the University of Oxford put it, when responding to the Commons inquiry report: "There was an assumption within the climate science community that we could use our professional judgment to distinguish between professional scientists and activists or members of the public. The big implication in all this for science is that the [Freedom Of Information Act] is taking away our liberty to use our own judgment to decide who we spend time responding to. And that has a cost."

Many scientists recognise that, faced with the requirements of freedom of information legislation, something has to change. But they argue that there should be limits to openness, and that replication should be kept within the mainstream scientific community, with its rules and etiquette about peer-reviewed publication, and done without, as the British government's chief scientist John Beddington puts it, "demotivating" researchers like Jones.

There could be special arrangements for replicating analysis of large data sets such as those assembled by Jones. Stephen Schneider of Stanford University in California, the founder and editor of *Climate Change*, said in an email to CRU scientists and others in January 2009: "Our best way of dealing with this issue of replication is to have multiple independent author teams, with their own codes and data sets, publishing independent work on the same topics... That is how credible scientific replication should proceed."

But is that enough? Do non-scientists, or amateur scientists, or scientists who run politically charged blogs, have an equal right to share scientific data? Jones believes not. He sees little purpose in sharing data with people operating outside peer review. In

October 2009 he told Graham Haughton, a geographer at the University of Hull: "Science should be conducted through the peer review literature, as it has been for over 300 years. The peer review system is the safeguard science has developed to stop bad science being published."

But in an age of the internet and freedom of information laws, the pressures for more open access are growing. McIntyre says: "I don't think there should be any issue of drawing up special rules for outsiders. I simply ask that scientists live up to their own policies. There is an unseemliness about scientists willingly providing data to their friends and resisting the provision of data to people who are perceived as critics." That was the standoff that led to escalating freedom of information requests at CRU through 2009. That, ultimately, was what triggered the bonfire of scientific vanities that followed the release of CRU's emails.

In the wake of Climategate, a few scientists are trying to chart a new path, to create links with the climate sceptics and with the wider challenge to the priesthood of science. The most forthright of these is Judy Curry of the Georgia Institute of Technology. She cut her teeth in public debate after publishing a paper linking climate change and hurricanes shortly after Katrina hit New Orleans, and reaped a whirlwind of vitriol for her pains. "I and my colleagues were totally bewildered and overwhelmed by the assault we found ourselves under" from sceptics, she says.

Initially, she sought refuge among scientists who had experience of such whirlwinds. "Associating with a tribe where others were more experienced and savvy about how to deal with this was a relief and very helpful at the time." They were able, she says, "to circle the wagons and point the guns outward in an attempt to discredit misinformation." But she has since attempted to break out from that tribal mentality, "trying to figure out how to engage sceptics effectively". Some called her a turncoat, particularly after she congratulated McIntyre for his work on freeing up data. In an open letter to young scientists involved in climate research, she said: "I have received significant heat from some colleagues for doing this. I've been told that I am legitimising the sceptics and misleading my students."

The publication of the CRU emails, she says, has reinforced her fears about a closed world of climate science – "the systematic and continuing behaviour from scientists that hold editorial positions, serve on important boards and committees and participate in major assessment reports", as she puts it. This encourages bad practice.

Reviewing the saga of the leaked emails in a series of blogs during 2010, she said the various data sets connected to Mann's hockey stick studies and Jones's CRU temperature record "stand out as lacking transparency". The raw data behind their key graphs "were not preserved" by the analysts, she said, though it "presumably is available from the original sources". Rather than being stuck in the archives, it needs reprocessing and reanalysing, she believes. And not just by mainstream scientists. She points out that it was McIntyre and McKitrick, rather than peer reviewers, who exposed poor statistical methods at work in the hockey stick. "Ignoring sceptics from outside the field is inappropriate. Einstein did not start his research career at Princeton, but rather at a patent office." The lesson may be: bring on the bloggers. There may be an Einstein among them.

She doesn't blame any individuals for this state of affairs. As Jones told the Commons inquiry, what he did was "standard practice". But "given the growing policy relevance of climate data, increasingly higher standards must be applied," she says. Mainstream scientists, she admits, "claim they would never get any research done if they had to continuously respond to sceptics. The counter to that argument is to make all of your data, metadata and code openly available. Doing this would keep molehills from growing into mountains."

Many climate scientists still don't get it, says Curry. "The mainstream IPCC types don't like what I'm doing at all." As a result, they misread the public debate after Climategate, she believes. "The climate research establishment appealed to its own authority and failed to understand that Climategate is primarily a crisis of trust." Along with Glaciergate and the travails of the IPCC, it exposed "a host of concerns that had been festering in the background". Those concerns included scientists acting as advocates, "tribalism that excluded sceptics", hubris over the IPCC's Nobel Prize, and "inattention to the statistics of uncertainty and complexity of alternative interpretations... Climategate isn't going to go away until these issues are resolved."

Tribalism meant her colleagues didn't see how the sceptical community had changed. In an interview in March 2010, Mike Mann summed up the prevailing view of what amounted to a corporate conspiracy against science. "Our detractors are extremely well funded, they are extremely well organised. They have basically had an attack infrastructure of this sort for decades," he said. "They developed it during the tobacco wars. They honed it further in their efforts to attack science that industry or other special interests find inconvenient."

That may once have been true, says Curry. An older generation of critics, most exemplified by Pat Michaels, had well-developed financial and intellectual links to the fossil fuel industry and its lobby groups. They fitted the stereotype of foot soldiers in a "denial machine funded by big oil," she says. But there is a new generation of "climate auditors, started by Steve McIntyre". They "have no apparent political agenda, are doing this work for free, and have been playing a watchdog role, which has engendered the trust of a large segment of the population." The public and some in the media had noticed this change. But many climate scientists had not.

Mann is not convinced about this new generation of freedom fighters for open science. "I think what is happening is the anti-science machine industry has fully exploited the resources made available by the world wide web. I would imagine that much of what might appear to an outsider to be organic, to be grassroots, is actually connected, funded, manned by those connected with the climate change denial movement." If it turns out that Climate-gate was the result of a corporately funded hack, he may yet be proved right.

Other leading figures besides Curry are looking for ways to rebuild trust in climate science. In the *Wall Street Journal* in December 2009, Hans von Storch, the German meteorologist and veteran of skirmishes with Mann and others, denounced the "CRU cartel" for their efforts to suppress open access to data. But, he went on: "We need to repair the damage and heal the public's new mistrust of the workings of climate science... The core of the knowledge about man-made climate change is simple and hard to contest. [Nonetheless] data must be accessible to adversaries; joint efforts are needed to agree on test procedures to validate, once again, already broadly accepted insights."

Storch advised that "the authors of the damaging emails would be wise to stand back from positions as reviewers and participants in the IPCC process. The journals *Nature* and *Science* must review their quality-control measures and selection criteria for papers." Meanwhile, he told politicians: "You have the knowledge you need for the political decisions. Let us [scientists] sit in our studies and discuss the remaining issues... Give us time to consider, to test alternative hypotheses, to falsify theories – to do our work without worrying if the results support our causes. Science is a valuable and unique societal institution, but not if it is consumed by short-sighted political goals."

One of Jones's former senior colleagues, ex-CRU research scientist Mike Hulme, joined with Oxford science philosopher Jerry Ravetz to write about trust in the aftermath of Climategate: "Climate scientists will have to work harder to earn the warranted trust of the public – and maybe that is no bad thing," they said. But to do that, science itself might have to change. "This event might signal a crack that allows for processes of restructuring scientific knowledge of climate change. It is possible that some areas of climate science have become sclerotic... too partisan, too centralised. The tribalism that some of the leaked emails display is something more usually associated with... primitive cultures." Ouch.

Ravetz told the *Guardian*: "In retrospect, it is clear that the Norwich group [CRU] were practising evangelical science. For them there was a simple truth that would save us, and all naysayers were evil." After the war on drugs and the war on terror, we now have a war on carbon. But this is based on a bogus certainty. We live, he says, in a world of "post-normal science, where facts are uncertain, values in dispute and decisions urgent." He called the "extended peer community" on the blogosphere "necessary for the health of science. In spite of all the hazards of any extension of democracy, the rejuvenating effects must be good."

And, whether it is necessary for science or not, he said that in the wake of Climategate, the power of this new force is probably unstoppable. "It is hard to see how this extended peer community of the scientific blogosphere could be silenced or suppressed, once it has shown its power. Doubtless it will be vulnerable to misuse and abuse, just like democracy in the political sphere, and so it will need guidance... and courtesy."

Courtesy. Maybe that is the toughest call of all.

Ravetz's remarks have been circulating through the British science community and generating a deal of comment, though so far as I can see the contribution from the climate scientists involved in the emails has been minimal. However, there are signs that the trauma of Climategate is leading to greater openness about research data.

On 27 November 2009, less than a week after the leak, Real-Climate set up a page on its site to provide "links to sources of temperature and other climate data, codes to process it, model outputs, model codes, reconstructions, paleo-records, the codes involved in reconstructions etc". The page is already a valuable resource. "The climate science community fully understands how important it is that data sources are made as open and transparent as possible, for research purposes as well as for other interested parties," the editors said. That is a sea-change from the days in the CRU bunker.

Britain's Met Office also announced that it was putting into the public domain data on climate change from 1,700 stations round the world. This was not as big a deal as it sounded. Jones had told *Nature* he was working on this back in July 2009. And, as the Met Office spokesman David Britton admitted, a lot of it was old data already "publicly available" through the World Meteorological Organisation.

The Met Office likes to pretend that it is foreigners who insist on restricting access to their data. But in fact, as Jones pointed out to me in the summer of 2009, one of the tightest confidentiality agreements signed by CRU covers data from the Met Office itself. The government agency regards the data as commercially valuable. Britton confirmed: "We at the Met Office have to offset our costs for the benefit of the taxpayer, so we have to balance that against freedom of access." And the Oxburgh report in April 2010 took time out from analysing the world of CRU to note that the government policy of trying to charge for meteorological data was "unfortunate and seems inconsistent with policies of open access to data promoted elsewhere in government". Moreover, it suggested that after the British began charging, "other countries have followed suit". Far from foreigners insisting on erecting walls against sharing their data, and Brits doing their best to maintain access, Oxburgh appears to conclude that the Brits started it.

The scientific journals also have an important role to play in improving access to data and codes. They can rule on how much material is put into the public domain before papers are approved for publication, for instance. Right now different journals have different policies. The big two, *Nature* in the UK and *Science* in the US, are relatively relaxed. They usually demand little material on top of what is required to allow the paper to pass muster with reviewers. And, as Jones revealed in cross-questioning before the House of Commons Climategate inquiry, that is generally not much.

Others are tougher. The Royal Society in London requires full data disclosure from contributors to its *Philosophical Transactions*. As we saw earlier, its insistence resulted in the release of Keith Briffa's Yamal tree ring data sets that had long been demanded by McIntyre. Some journals are reassessing their positions. Schneider, founder and editor of *Climate Change*, asked his editorial board to consider the matter after McIntyre asked him for some of his personal computer codes. The board, whose members include Jones, decided that enough data should be provided to allow other researchers with the necessary skills to write their own codes in order to replicate the findings. But no more.

Schneider thinks that may not be the end of the matter. He told the *Guardian* there might be some middle ground between full disclosure and the desire of researchers to derive benefits from personally written computer codes. Maybe, like commercial patents, they should be allowed exclusive use of their own codes, as their personal intellectual property, for two or three years. That would be time enough to "publish the initial papers using their hard work". But after that, the codes should all be disclosed. He added: "This broad discussion about the boundaries of data transparency, personal codes and exclusive rights… may be the only positive that might emerge from this unfortunate incident."

Many sceptics are not satisfied with such half-way houses. They are in full cry against the entire peer review process. Some talk about "peer-to-peer" review, meaning an end to centralised control through journals and a free-for-all in which everything is published and anyone can comment on anything. The west coast journalist Patrick Courrielche, who is active in this movement, claims that Climategate "triggered the death of unconditional trust in the scientific peer review process, and the maturing of a new movement of peer-to-peer review".

Can an entirely free intellectual market deliver better science? Can the pioneers of scientific review on the blogosphere do better than the journals? Would this ensure quality control or shatter it? Should the Jeffrey Archers of the scientific world have as much access to the journals as the Pulitzer prize-winners? They may shudder in the labs, but one day we may find out. For like other recent battles over access to publicly held information, from lists of paedophiles to school league tables, those demanding freedom of information are winning the argument.

CHAPTER 19

Saving the science

"Climate scientists will have to work harder to earn the warranted trust of the public – and maybe that is no bad thing"

MIKE HULME ON THE FUTURE, DECEMBER 2009

Is the science of climate change fatally flawed? Have the Climate-gate revelations undermined the case that we are experiencing man-made climate change? Absolutely not. Nothing uncovered in the emails destroys the argument that humans are warming the planet. None of the 1,073 emails, or the 3,587 files containing documents, raw data and computer code upsets the 200-year-old science behind the "greenhouse effect". We might wish it weren't so, but the world still has a problem. A big problem.

First the basic physics. It is beyond doubt that certain gases in the atmosphere, most importantly water vapour and carbon dioxide, trap heat. They have a "greenhouse effect". This in itself is no bad thing. Without those gases the planet would freeze. There is also no doubt that human activity – mostly burning carbon-based fuels like coal and oil, but also deforestation and ploughing – is pumping carbon dioxide into the atmosphere. Nor do even the most die-hard sceptics deny that this has caused a sustained year-on-year rise in carbon dioxide concentrations. For almost 60 years, continuous measurements at the Mauna Loa observatory in Hawaii have charted this rise. Today's concentrations are about 35 per cent above pre-industrial levels.

Knowing what we know about those gases, we would expect them to warm the atmosphere, and we would expect the atmosphere in turn to warm the oceans, starting at the surface. All these things we see happening.

In 2000, scientists for the first time measured the heat trap directly. Researchers at Imperial College London examined satellite data covering almost three decades to plot changes in the amount of infrared radiation (heat) escaping from the atmosphere into space. They looked in particular at the part of the infrared spectrum that is trapped by carbon dioxide – wavelengths between 13 and 19 micrometres. They found that between 1970 and 1997 less and less radiation at these wavelengths was escaping.

But this greenhouse effect is not the only issue. On its own, a doubling of carbon dioxide levels in the atmosphere would raise global temperatures by an average of only about 1C. No big deal. But all the evidence is that this initial warming will be amplified, probably about threefold, by "feedbacks". And here legitimate debates begin about the extent and nature of the feedbacks.

The most obvious of them is the melting of ice, which covers significant parts of the earth's surface. Ice is a superb reflector of solar energy. It bounces right back into space. Less ice means more solar radiation is absorbed by the dark surfaces of the oceans and land, and radiated back into the air as infrared heat.

Probably about as important, but much less certain, is the impact of a build-up of water vapour in the atmosphere. This happens because a warmer world means more water is evaporated from the oceans. Water vapour is a potent greenhouse gas, so more water vapour adds to the heating. How much is difficult to be sure of, since it depends on how much of that water vapour turns into clouds (which shade the earth in the day, but keep it warm at night), and on how fast those clouds "rain out" and disappear.

The sceptical meteorologist Dick Lindzen has argued that changes to clouds could end up making the atmosphere drier rather than wetter, which would be a negative feedback, counteracting the greenhouse warming. So far, there seems to be no evidence of this in practice, but it is possible. Another human influence is the dust and aerosol pollution that persuaded climate scientologists in the 1970s that we could be destined for cooling. The science is now established that it simply takes the edge off global warming, but again the numbers are debatable. In the long run this won't help us much, because while these pollutants generally stick around in the air for only a few days, greenhouse gases like carbon dioxide accumulate. The greenhouse effect will win in the end.

But such questions lie behind the continuing uncertainty among all scientists about exactly how much the world will warm, and how quickly. The IPCC reckons that on current trends, we can expect between 2C and 4C of warming this century. Even arch-sceptics like Lindzen, Michaels and McIntyre agree the world will warm. The only argument is about how much – and how much we need to worry.

An obvious way to help us figure out how much we should be concerned is to try to establish our influence so far – how much we have warmed our world through the recent accumulation of gases like carbon dioxide. Attention has focused on the warming of the final decades of the 20th century. Past climate fluctuations were due to natural forces, so can they explain this recent sharp warming as well?

The most obviously influence is the sun. Probably until the mid-20th century fluctuations in solar radiation were the main influence on global temperatures, either because of changes in the sun or because of wobbles in the earth's orbit, which seem to trigger the regular ice ages. The trouble is that since about 1980, solar influences ought to have been cooling the planet. They could not have been causing warming. Another known influence is volcanic eruptions, which can shade the Earth with sulphate particles that cause cooling. If we were recovering from a period of strong volcanic activity, that might explain the warming. But again that does not fit the evidence from the real world.

There is, however, one final element that has been rather ignored until recently: the influence of the oceans. Most heat at the planet's surface ends up in the oceans. That includes the additional heat being trapped by greenhouse gases. But it is becoming clear that the balance between how much heat is in the oceans and how much is in the atmosphere varies. Climate researchers have long known that during the periodic fluctuations in the Pacific Ocean known as El Niños, the atmosphere warms. This is because, during El Niños, a layer of very warm water spreads across the equatorial Pacific and warms the air. El Niños only last for a year or so, but it now seems that other long-term oceanic oscillations, acting over decades in the Pacific and Atlantic, could have similar influences over longer periods – either extracting extra heat from the atmosphere or releasing it into the atmosphere.

Such ocean oscillations are now thought to have been an element in the anomalous atmospheric warming in the 1930s and 1940s, which Jones acknowledged in early 2010 had been as marked in intensity (though not in length) as the warming in the late 20th century. And oceanographers say that both the Atlantic and Pacific oscillations were in warming phases in the final years of the 20th century. Mojib Latif of the University of Kiel told the World Climate Conference in September 2009 that part of the warming, particularly in the North Atlantic and Arctic regions since the 1970s may be due to such an influence. "But how much? The jury is still out," he said. Since 2002, the Pacific has been in a cooling phase, and many believe that has helped cause the levelling-off of global temperatures since then. Nobody is sure.

I first investigated these new analyses in 2008. I wrote in the *Guardian*: "We can expect the deniers to make the most of this opportunity to pour cold water on the whole climate change narrative. No year has yet been hotter than 1998, they will say. True: it was a huge El Niño year. Now we are on the way back down, they will say. Nonsense. The underlying trend remains upwards; and as every decade passes, natural cycles can do less and less to counter the growing human influence on temperature."

Judy Curry has accused some climate scientists of trying to airbrush these 20th century ocean oscillations from their analysis, rather than trying to understand them. Certainly, when I reported the cooling prognosis for the next few years, I met some hostility from climate scientists. But other researchers have taken up the challenge. A determination to get to the bottom of this issue was what drove Kevin Trenberth to make his much-misconstrued remark about it being a "travesty" that the recent temperature trends were not yet explained. Far from trying to cover it up, he wanted to investigate.

Even people who believe the ocean influences are large accept that they are transient, lasting a couple of decades at most, whereas accumulating carbon dioxide is not. By late next decade, natural warming will once again combine with man-made warming to push temperature rise into overdrive. The surge that we saw through the 1980s and 1990s will resume with a vengeance. That could be the moment that climate change passes a point of no return, when ice sheets start to collapse and parched rainforests and soils dump their carbon into the air, accelerating warming.

The truth is that the basic physics of the "greenhouse effect" is extremely robust and cannot fail to warm the planet. The question of the uniqueness or otherwise of 20th century warming is scientifically interesting, because it may help us understand the true sensitivity of the atmosphere to accumulating greenhouse gases. But beyond that it is a sideshow. The IPCC did the world no great favours by adopting the hockey stick as a totem of global warming. We know the planet has been warmer and colder in the past through entirely natural forces. It does not matter if it was as warm during the medieval period as today.

How will things develop? Ocean oscillations will continue to cause ripples. The rising greenhouse gas concentrations will become ever more dominant. But to what ultimate end? Here the uncertainties remain considerable. Conventional climate models cannot say for certain whether a doubling of CO_2 in the air will raise temperatures by 1C or 5C, though 3C looks a reasonable expectation, unless some unexpected feedbacks kick in.

Such feedbacks could be negative, as Lindzen suggests. But much more likely, in most estimations, are positive feedbacks. All the evidence of climate history is that periods of stability are punctuated by sudden temperature jumps driven by such feedbacks. These periods of abrupt change are usually triggered by solar influences but then amplified and accelerated by carbon dioxide – the very stuff we are putting into the atmosphere with such abandon today.

There are two main concerns today. One is that warming will release greenhouse gases from natural carbon stores – carbon dioxide from dying rainforests or parched soils; methane from melting permafrost or frozen stores known to exist in very large volumes beneath the ocean floor. The second is that the accelerated break-up of ice sheets will cause much more rapid sea level rise than currently envisaged by glaciologists. Both concerns were raised in the last IPCC report but, as we have seen, they were largely kept out of the summary for policy-makers.

It seems the climate sceptics are right about one thing. They are right to say we know less than we think about how climate change will play out. Where they are wrong, and illogical, is in presuming that our uncertainty means we can worry less. In my judgment, we should worry a lot more. After all, when we put carbon dioxide into the atmosphere we are playing with what looks like the planet's thermostat.

Climategate may not have changed the "big picture" of climate change. But the emails have certainly raised new concerns about the reliability of the painstakingly assembled temperature record. And they have done significant damage to the wider credibility of climate scientists. As Oxburgh concluded, poor record-keeping is a cardinal sin when "the potential for misleading results arising from selection bias is very great". And poor record-keeping was, on its own admissions, a continuing failure within CRU over many years.

The evidence of scientists cutting corners, playing down uncertainties in their calculations, and then covering their tracks by being secretive with data and suppressing dissent suggests a systemic problem of scientific sloppiness, collusion and endemic conflicts of interest, but not of outright fraud. Many scientists believe they have to own up to that, and put their house in order.

They also have to change how they work. The doors of the labs are being opened, whether scientists like it or not. The science community now urgently needs to figure out how to respond to this altered landscape, and to stop fearing change. Scientists were among the first to make serious use of the internet, once it ceased to be a medium of military communication a generation ago. I remember watching in awe back then as scientists at CRU showed me how they could instantly download large data sets from a computer on the other side of the Atlantic. Yet now the rest of us have that capability, they seem to fear its power. They have to embrace that change, recognise that it can extend access to data and open up peer review. The internet once helped build their international cabals of collective interest; now it needs to be employed to break them down.

Finally, Climategate raises questions about the IPCC report-writing process, in which many of the emailers have been involved. Governments set up the IPCC more than 20 years ago to get scientists to speak with one voice on climate change. But often there is no clear consensus. Scientists are trained to disagree, but they have sometimes become drawn into what amounts to a polit-ical process of cobbling together bogus consensus. That is both bad politics and bad science.

The drive to find agreement where there is none has created pressure to keep the message simple and for scientists who have a problem with that to keep quiet. Some shut up. Others bend their results or curtail their researches to fit the prevailing view, arguably slowing the process of scientific discovery. Others react with anger and end up among the outright sceptics. Such tensions are clear in dozens of the CRU emails. They created the two tribes.

Another concern of many is that, though the IPCC was set up in 1988, four years before the first UN agreement on climate change, it has gradually become a creature of the political imper-atives of the UN negotiating process. "The questions posed at the IPCC have been framed in terms of [that] agenda," says Curry.

"That's caused a narrowing of the kind of things the IPCC focuses on. It's not a policy-free assessment of the science."

Remedying such failings may require an end to the IPCC in its present form, and certainly reforms much wider than those currently on the table. Jones's former colleague at CRU, Mike Hulme, says that "through its structural tendency to politicise climate change science [the IPCC] has perhaps helped to foster a more authoritarian and exclusive form of knowledge production – just at a time when globalising and wired cosmopolitan culture is demanding of science something much more open and inclusive."

In December 2009, meteorologist Roger Pielke Sr criticised the "broad goal of the leadership of the IPCC process to control what science the policy-makers receive". The leaked emails certainly expose that tendency. But the trouble is that the IPCC was set up by governments to do precisely that – to be a gatekeeper for the creation of scientific consensus. Blaming the scientists is too easy. Climategate raises important questions about what we want of our scientists, in the field of climate and elsewhere.

NOTES ON SOURCES

All of the Climatic Research Unit (CRU) emails published online are available in a searchable form at www.eastangliaemails.com. Below I have given them the reference code found on that site, eg "933255789.txt".

The House of Commons report, along with the transcribed oral hearings and submissions to it, is published at www.publications. parliament.uk/pa/cm200910/cmselect/cmsctech/387/387i.pdf andwww.publications.parliament.uk/pa/cm200910/cmselect/ cmsctech/387/387ii.pdf.

The Muir Russell inquiry website is at www.cce-review.org.

The CRU website is at www.cru.uea.ac.uk.

The annotated versions of my original online reports for the *Guardian* are at www.guardian.co.uk/environment/series/ climate-wars-hacked-emails.

Introduction: Windows on a closed world

My first articles on the hack in *New Scientist* appeared as "Hacked archive provides fodder for climate sceptics" and "Stay on the moral high ground", both 28 November 2009; and "Stand and deliver your climate data", 19 December 2009. See also my article at Yale360 at www.e360.yale.edu/content/ feature.msp?id=2221.

My 1996 tree rings article was "Lure of the rings", *New Scientist*, 14 December 1996. The emailed draft is here: 845217169.txt. The bloggers' take is here: http://baseballmedia watch.blogspot.com/2009/12/biblical-intensity-in-search-for-sign.html.

Chapter 1: Two tribes go to war

Santer's "dark alley" remark is at 1177534709.txt. Mann's "dirty laundry" remark is at 1059664704.txt.

Campbell's departure is announced at www.cce-review.org/News.php and http://bishophill.squarespace.com/blog/2010/2/11/campbell-resigns.html.

McIntyre describes revisited his mining career here: http://climateaudit.org/page/2/. McKitrick's evidence to the Commons is here: www.publications.parliament.uk/pa/cm 200910/ cmselect/cmsctech/387/387ii.pdf.

Curry speaks to *Discover* on M&M here: http://discover-magazine.com/2010/apr/10-it.s-gettin-hot-in-here-big-battle-over-climate-science. And Santer's responses on RealClimate are here: www.realclimate.org/index.php/archives/2010/02/close-encounters-of-the-absurd-kind/; and in an email to Karl here: 1226451442.txt.

RealClimate's complaint about "mini-blogstorms" is here: www.realclimate.org/index.php/archives/2009/09/hey-ya-mal/.

McKitrick's comments about the IPCC are here: http://network.nationalpost.com/NP/blogs/fullcomment/archive/2009/10/01/ross-mckitrick-defects-in-key-climate-data-are-uncovered.aspx.

Chapter 2: In the freezer

Schneider's 1971 paper is "Atmospheric carbon dioxide and

aerosols: Effects of large increases on global climate", *Science*, v173, p138. He revisited it in "Climate stabilisation: For better or for worse?" with Kellogg in *Science*, v 186, p1163. Thomas Petersen wrote on "The myth of the 1970s global cooling scientific consensus" in the *Bulletin of the American Meteorological Society*, September 2008, p1325. Lenton's 2007 paper appeared in *Climate Dynamics*, v 29, p591.

Lamb's 1965 study is in *Palaeogeography, Palaeoclimatology, Palaeoecology*, v 1, p13. And his BA lecture, given in Norwich in 1984, was called "The future of the earth: greenhouse or refrigerator?" Kelly's obituary of Lamb is in *Nature*, v 388, p836.

Some of the material in this chapter comes from my article in *New Scientist*, "The ice age that never was", 15 December 2006.

Chapter 3: The secret origins of the IPCC

Much of the background scientific material and the concluding statement from the Villach conference appear in *The Greenhouse Effect, Climate Change and Ecosystems*, ed Bolin, Doos, Warrick and Jaeger, published on behalf of SCOPE (Scientific Committee on Problems of the Environment), by John Wiley, Chichester, in 1986. www.icsu-scope.org/downloadpubs/scope29/preface.html.

Much of the analysis here and some of the quotes come from a report by Wendy Franz to Jaeger at IIASA, *The development of an international agenda for climate change: Connecting science to policy*, available here: www.iiasa.ac.at/Admin/PUB/Documents/IR-97-034.ps.

The Brundtland Report (properly *Our Common Future*, Oxford UP, 1987) is available here: www.un-documents.net/wced-ocf.htm. Read the *New York Times's* contemporary report on Hansen in the Senate at www.nytimes.com/1988/06/24/us/global-warming-has-begun-expert-tells-senate.html.

The statement from the World Conference on the Changing Atmosphere: Implications for Global Security, in Toronto, is here: www.cmos.ca/ChangingAtmosphere1988e.pdf.

Chapter 4: Playing hockey with climates past

My article on Briffa's work, "Lure of the rings", appeared in *New Scientist*, 14 December 1996.

Mann *et al*'s first paper on the hockey stick, "Global-scale temperature patterns and climate forcing over the past six centuries", appeared in *Nature*, v 392, p779. The second paper, "Northern hemisphere temperatures during the past millennium: Inferences, uncertainties and limitations" was in *Geophysical Research Letters*, v 26, p759. Jacoby is quoted from my article "The great hockey stick debate", *New Scientist*, 18 March 2006.

See the hockey stick in the IPCC 2001 report here: www.grida.no/publications/other/ipcc_tar/. Both Folland's email about the "clear favourite" and Briffa's response are here: 938018124.txt

Briffa and Osborn's *Science* paper is "Seeing the wood from the trees", v 284, p926.

Bradley's dissociation is at 924532891.txt. And the Jones attack on Mann is here: 926026654.txt.

The "completely separate" quote is from a RealClimate response to the *Guardian* articles here: www.realclimate.org/index.php/archives/2010/02/the-guardian-disappoints/.

Chapter 5: The Yamal tree ring "virus"

The first email was this: 826209667.txt. My draft article on Briffa's work is here: 845217169.txt. Briffa and Shiyatov wrote in *Nature*, "Unusual twentieth century summer warmth in a thousand-year temperature record from Siberia", v 376, p156.

Davies discussed fraud at 1256214796.txt. McIntyre discussed "the most important tree" at http://climateaudit.org/2009/09/30/yamal-the-forest-and-the-trees/.

The Briffa *et al Philosophical Transactions* paper is "Trends in recent temperature and radial tree growth spanning 2,000 years across northwest Eurasia", *Philosophical Transactions of the Royal Society B-Biological Sciences* v 363, p2271.

McIntyre admits having the Russian data here: http://climateaudit.org/2009/10/05/yamal-and-ipcc-ar4review-comments/#comment-197561; and says much more about Yamal here: http://climateaudit.org/category/proxies/yamal-and-urals/.

Kaminsky speaks up here: http://spectator.org/archives/2009/12/04/dead-ringer; and McKitrick here: http://wattsup withthat.com/2009/10/02/ross-mckitrick-sums-up-the-yamal-tree-ring-affair-in-the-financial-post/.

Jones reiterates his support for Briffa on Yamal in his BBC interview: http://news.bbc.co.uk/1/hi/8511670.stm.

Briffa defends himself here: www.cru.uea.ac.uk/cru/people/briffa/yamal2009/ and here: www.cru.uea.ac.uk/cru/people/briffa/yamal2000/.

But Wigley on Briffa is here: 1254756944.txt

And Briffa's 1999 *Science* warning, "Seeing the wood from the trees", is in v 284: pp926-927.

RealClimate lays into hockey stick deniers here: www.real climate.org/index.php/archives/2009/09/hey-ya-mal/.

Delingpole goes over the top at http://blogs.telegraph.co.uk/news/jamesdelingpole/100011716/how-the-global-warming-industry-is-based-on-one-massive-lie/.

McIntyre's message to the *American Spectator* turns up here: 845217169.txt.

Chapter 6: Urban heat and a Chinese riddle

Jones's 1990 paper, "Assessment of urbanisation effects in time series of surface air temperatures over land" is in *Nature*, v 347 p169. The IPCC chapter three is here: www.ipcc.ch/publications _and_data/ar4/wg1/en/ch3.html.

The Jones and Yan 2000 paper is in *Advances in Atmospheric Sciences*, v 18, p309.

Jones's "marked man" email is at 1177158252.txt, where Mann advises on "charlatans" and Trenberth urges a fightback. Jones's "my problem" email is at 1182346299.txt.

Keenan has posted a letter from the University of Albany advising him about its findings on Wang at www.informath.org/ apprise/a5620/b523.pdf. He has posted his paper in *Energy and Environment* (v 18, p985) alleging fraud at www.informath.org/ pubs/EnE07a.pdf, along with Wang's 1990 paper, "Urban heat islands in China", from *Geophysical Research Letters* (v 17, p2377), at www.informath.org/apprise/a5620/b23.pdf.

The exchange between Jones and Wang over what to do about Keenan is at 1188508827.txt. Mann weighs in at 1188478901.txt. Peiser says he will publish anyway at 1189515774.txt.

Wigley's two emails to Jones about Wang are at 1188557698.txt and 1241415427.txt.

Later, Jones speaks to *Nature* at www.nature.com/news/ 2010/100217/full/463860a.html. His 2008 paper, "Urbanisation effects in large-scale temperature records, with an emphasis on China", is in the *Journal of Geophysical Research*, v 113, pD16122. Parker's assessment of the "Urban heat island effects on estimates of observed climate change" is in electronic form at *Wiley Interdisciplinary Review: Climate Change* (DOI 10.1002/ wcc.21)

Chapter 7: Scientists in the firing line

Wigley on Lindzen is at 1106338806.txt. Michaels's blog is at
www.worldclimatereport.com. His *New Scientist* article is "Global
pollution's silver lining", 23 November 1991. I reported on him,
Lindzen and others in *New Scientist* in "Greenhouse wars", 19
July 1997. The *New York Times* reported his link to Western Fuels
on 27 July 2006: "Utilities pay scientist ally on warming".

The CEI's commercial for CO_2 is at www.youtube.com/
watch?v=7sGKvDNdJNA.

I wrote about the Global Climate Science Communications
Plan in "Experts with a price on their heads", *Guardian*, 7 May
1998. My piece on preparing for the IPCC appeared in *New Scientist*, "State of denial", 4 November 2006. And my article on
hurricanes and climate change, "Is global warming making hurricanes stronger?", in *New Scientist*, 3 December 2005.

Curry's 2005 paper on hurricanes, "Changes in tropical cyclone
number, duration, and intensity in a warming environment" by
Webster *et al* appeared in *Science*, v 209, p1844. RealClimate's
comment was at www.realclimate.org/index.php/archives/2005/
09/hurricanes-and-global-warming/. Trenberth gave his pre-
Katrina interview to *Nature*, v 435, p1008. Michaels responded at
www.worldclimatereport.com/index.php/2005/06/24/a-man-
on-a-mission/.

Knutson's paper is "Simulated reduction in Atlantic hurricane
frequency under twenty-first-century warming conditions",
Nature Geoscience, v 1, p359.

Chapter 8: Breaking the hockey stick

Inhofe's 2005 senate statement is here: www.govtrack.us/
congress/record.xpd?id=109-s20050104-25. *Nature*'s news item
on Barton is "Climate of distrust", v 436, p1. Crowley's comments
are at www.geo.umass.edu/climate/crowleyeos.pdf.

Broecker is quoted from my article "The great hockey stick debate", *New Scientist*, 18 March 2006; and Curry from her *Discover* interview: http://discovermagazine.com/2010/apr/10-it.s-gettin-hot-in-here-big-battle-over-climate-science.

The M&M paper in *Energy and Environment* is at www.multi-science.co.uk/mcintyre-mckitrick.pdf. Mann's suggested response is at 1067194064.txt. But his final correction to the *Nature* paper of 1998 appeared in *Nature*, v 430, p105.

M&M's paper, "Hockey sticks, principal components, and spurious significance", appeared in *Geophysical Research Letters*, v 32, pL03710.

The *National Post* article "Research debunks Kyoto hockey stick" can be read here: www.anticorruption.ca/forum/ phpBB2/ viewtopic.php?t=98&sid=adb16526580c6eb446e9839f78036876. Wigley's warning is at 1098472400.txt. Mann's vitriol is at 1067596623.txt. Jones's response to Wigley is at 1098472400.txt.

Hans von Storch's 2004 review, "Reconstructing past climate from noisy data", was in *Science* v 306 p679. The Wahl and Ammann's paper, "The importance of the geophysical context in statistical evaluations of climate reconstruction procedures", is in *Climate Change*, v 85, p71.

McKitrick's article in the *Financial Post* is at: http://network. nationalpost.com/np/blogs/fullcomment/archive/2009/10/ 01/ross-mckitrick-defects-in-key-climate-data-are-uncovered.aspx.

Jones's "keep them quiet" email is at 1116611126.txt. Briffa avoids over-egging at 1140039406.txt. The National Academy of Sciences' findings on the hockey stick are at http://books.nap. edu/openbook.php?record_id=11676&page=R1. And Cuffey's ideas are recorded here: www.nature.com/nature/journal/ v441/n7097/full/4411032a.html.

Chapter 9: Discerning the hand of man

Read the key part of the 2001 IPCC report here: http://env.chass.utoronto.ca/env200y/ESSAY2001/science. htm#four. Claimed changes to the IPCC text are discussed here: www.sepp.org/Archive/controv/ipcccont/Item03.htm.

Santer's key paper, "A search for human influences on the thermal structure of the atmosphere", was in *Nature*, v 382, p39. Later he battled it out with Michaels in *Nature*'s letters pages, v 384, p524. Seitz's *Wall Street Journal* piece is here: www.sepp.org/Archive/controv/ipcccont/Item05.htm.

The Oregon Petition Project can now be viewed here: www.petitionproject.org.

Santer's comments on the *Guardian* are here: www.realclimate.org/index.php/archives/2010/02/close-encounters-of-the-absurd-kind/. The "scientific cleansing" charge is made in "The IPCC: Institutionalised 'scientific cleansing' ", Global Climate Coalition memorandum, Donald Rheem, May 1996.

I interviewed Santer about this for *New Scientist* in "Where angels fear to tread", 4 March 2000.

Chapter 10: Satellites, stats and bogeyman Santer

Santer's paper, "Consistency of modelled and observed temperature trends in the tropical troposphere", is in the *International Journal of Climatology*, v 28, p1703. Douglas and Christy's paper is here: www.pas.rochester.edu/~douglass/papers/Published%20JOC1651.pdf.

Jones discussed events in April 2007 here: 1196795844.txt. Santer talked of "intellectual dishonesty" here: 1199988028.txt. Osborn was on the case here: 1200076878.txt. And reported back here: 1200076878.txt.

Douglass riposted in the *American Thinker* here: www. amer-icanthinker.com/2009/12/a_climatology_conspiracy.html. Santer responds to McIntyre on replicating his work at 1226337052.txt. McIntyre blogs in response at http://climateaudit.org/2008/11/10/santer-refuses-data-request/.

Santer discussed with Karl at 1226500291.txt; and later with Wigley at 1228330629.txt; revealed problems with his bosses here: 1228249747.txt; and changed his mind here: 1228330629.txt. Wigley said "Well done" here: 1229468467.txt.

Chapter 11: Silencing critics

Schmidt's remarks are from www.realclimate.org/index.php/archives/2010/02/the-guardian-disappoints/. Jones's email on rejecting two papers is at 1080742144.txt.

McIntyre's blog of CRU's policies is at http://climate audit.org/2009/12/21/climategatekeeping-siberia/.

Jones was interviewed by the BBC here: http://news.bbc.co.uk/1/hi/8511670.stm.

Briffa emailed Cook here: 1054748574.txt; and Cook replied here: 1054756929.txt.

The Soon and Baliunas paper, "Proxy climatic and environmental changes of the past 1,000 years", *Climate Research* (v 23, pp89-110), can be found at www.cfa.harvard.edu/~wsoon/1000 yrclimatehistory-d/Jan30-ClimateResearchpaper.pdf, and in its alternative version at www.marshall.org/pdf/materials/136.pdf. The Harvard press release is at www.cfa.harvard.edu/news/archive/pr0310.html.

Mann discusses a journal boycott with Jones in a series of emails at 1047388489.txt. And returns to the fray at 1057941657.txt.

Kinne changed his mind at: http://www.int-res.com/articles/misc/CREditorial.pdf.

Michaels's op-ed in the *Wall Street Journal* is reproduced by the Cato Institute at www.cato.org/pub_display.php?pub_id=11072. Storch's article is here: http://online.wsj.com/article/SB1000 14240527487042381045746014439470785538.html.

Jones's "garbage" email is at 1089318616.txt. The McKitrick and Michaels paper, "A test of corrections for extraneous signals in gridded surface temperature data", is in *Climate Research* (v 26, p159) and online at www.int-res.com/articles/cr2004/26/c026p159.pdf.

The IPCC chapter is here: www.ipcc.ch/publications _and_data/ar4/wg1/en/ch3s3-2-2-2.html. Mackwell from GRL is emailing at 1106322460.txt, and Mann and Wigley sum up here 1106322460.txt.

Chapter 12: Free the data

Jones spoke about doing things by the book to the BBC here: http://news.bbc.co.uk/1/hi/8511670.stm. He emailed Mann about avoiding FOI here: 1107454306.txt. To Wang here: 1182255717.txt. To Santer here: 1228330629.txt. And to Osborn here: 844968241.txt.

Osborn raised the confidentiality issue with Ammann here: 1212166714.txt. Ammann replied at 1212156886.txt. Jones's damaging requests to Mann and Bradley to delete emails appear here: 844968241.txt. The circumstances behind the ICO's statement are revealed here: http://climateaudit.org/2010/02/ 26/ hard-to-imagine-more-cogent-prima-facie-evidence/.

Jones wrote to Schmidt on exempting IPCC correspondence here: 1219239172.txt; and Briffa wrote to Ammann here: 1214228874.txt. The "ten pounds" quote comes here: 1228330629.txt.

Hughes tells his story here: www.warwickhughes.com/ blog/?p=207; and here: www.publications.parliament.uk/pa/cm 200910/cmselect/cmsctech/memo/climatedata/uc3502.htm, which includes Jones's 2005 email to him.

Jones to Wigley on IPR is here: 1106338806.txt. Schmidt wrote consolingly to Santer here: 1228258714.txt. And Schneider's round-robin is here: 1106338806.txt.

McIntyre's discussion of a mole appears at http://climate audit.org/2009/07/25/a-mole/. And http://climateaudit.org/ 2009/07/28/met-officecru-finds-the-mole/. *Nature* discusses the affair at www.nature.com/news/2009/090812/full/460787a. html. UEA describes requests as "vexatious" here: www.uea.ac. uk/mac/comm/media/press/2009/nov/CRUupdate.

McIntyre discusses UEA turning down his request here: http://camirror.wordpress.com/2009/11/21/test/.

Jones's prescient 2005 warning is at: 1107454306.txt.

Chapter 13: The search for the hacker
The *Guardian* inquiries into who did the hack were summarised on 4 February 2010 in "Climate emails: were they really hacked or just sitting in cyberspace?": www.guardian.co.uk/environment/ 2010/feb/04/climate-change-email-hacker-police-investigation.

An interesting analysis suggesting an "inside job", by Lance Levsen, appeared here: www.smalldeadanimals.com/FOIA _Leaked/.

Climategate: The Crutape Letters by Steven Mosher and Thomas W Fuller, ISBN 9781450512435, was privately published but is available online from Amazon etc.

RealClimate discusses the hack here: www.realclimate.org/ index.php/archives/2009/11/the-cru-hack/.

Patrick Courrielche's three interesting and well-informed articles start here: http://bigjournalism.com/pcourrielche/2010/01/08/peer-to-peer-review-how-climategate-marks-the-maturing-of-a-new-science-movement-part-i/.

McIntyre's version of events is here: http://climateaudit.org/2010/01/12/the-mosher-timeline/. Holland's "misconfiguration" story is here: http://climateaudit.org/2009/07/28/met-officecru-finds-the-mole/. And Charles the Moderator's intriguing take is here: http://wattsupwiththat.com/2009/11/23/the-crutape-letters%C2%AE-an-alternate-explanation/.

Chapter 14: Tricks and lies

May spoke here: www.desmogblog.com/elizabeth-may-informed-look-east-anglia-emails. Jones's "trick" email is at 942777075.txt.

Palin's *Post* piece is here: www.washingtonpost.com/wp-dyn/content/article/2009/12/08/AR2009120803402.html.

Inhofe's Copenhagen speech is here: http://epw.senate.gov/public/index.cfm?FuseAction=Minority.Speeches&ContentRecod_id=9cac1e35-802a-23ad-4540-3e4706eab1bd&Region_id=&Issue_id=.

Trenberth's "travesty" message is here: 1255352257.txt; and his *Current Opinion* paper is here: www.cgd.ucar.edu/cas/Trenberth/trenberth.papers/EnergyDiagnostics09final2.pdf.

Michaels's misattribution is at www.cgd.ucar.edu/cas/Trenberth/trenberth.papers/EnergyDiagnostics09final2.pdf. See the original email at: 1089318616.txt.

See what Will was apparently quoting at: 1054736277.txt.

Chapter 15: A public disaster

UEA statements on Climategate appear here: www.uea.ac.uk. Santer defends Jones here: http://groups.google.com/group/

geoengineering/msg/8aab47682ab84454. Mann was quoted here: www.nytimes.com/cwire/2009/11/24/24climatewire-stolen-e-mails-sharpen-a-brawl-between-clima-19517.html. Ward was writing in the *Guardian* here: www.guardian.co.uk/environment/2009/nov/20/climate-sceptics-email-hacking.

The vile stuff from the Free Republic can be found at: www.freerepublic.com/focus/f-news/2447063/posts.

Delingpole's "nail in the coffin" piece is here: http://blogs. telegraph.co.uk/news/jamesdelingpole/100017393/climategate-the-final-nail-in-the-coffin-of-anthropogenic-global-warming/. Monbiot called on Jones to resign in the *Guardian* here: www.guardian.co.uk/commentisfree/cif-green2009/nov/23/global-warming-leaked-email-climate-scientists.

Mann is in the *Washington Post* here: www.washington post.com/wp-dyn/content/article/2009/12/17/AR200912 1703682.html, and in *Discover* here http://discovermagazine. com/2010/apr/10-it.s-gettin-hot-in-here-big-battle-over-climate science/article_view?searchterm=michael%20mann & b_start:int=2.

Weart is interviewed here: http://voices.washingtonpost. com/capitalweathergang/2009/11/perspective_on_a_climate_ scien.html. Lawson's foundation is at www.thegwpf.org.

The *Daily Express* story is here: www.express.co.uk/posts/view/146139/100-reasons-why-global-warming-is-natural. *New Scientist* replied here: www.newscientist.com/blogs/shortsharp science/2009/12/50-reasons-why-global-warming.html.

The *Guardian* reported on Inhofe's witch hunt here: www.guardian.co.uk/environment/2010/mar/01/inhofe-climate-mccarthyite.

Cyber-bullying is examined here: www.scientificamerican.com /article.cfm?id=climate-cyber-bullying&page=2. Monbiot added

Ogden to his "must go" list at www.guardian.co.uk/environment/georgemonbiot/2010/feb/02/climate-change-hacked-emails.

Jones was interviewed by *Nature* here: www.nature.com/news/2010/100215/full/news.2010.71.html.

Jo Abbess loves Jones here: www.joabbess.com/2010/02/10/we-love-you-phil-jones/.

May spoke here: www.desmogblog.com/elizabeth-may-informed-look-east-anglia-emails. Lean attacked the gutless at www.telegraph.co.uk/earth/earthcomment/geoffrey-lean/7273163/Green-activists-are-losing-their-fire.html.

Lynas attacked "barbarians at the gate" here: www.newstatesman.com/blogs/the-staggers/2010/02/lynas-climategate-scientists, while Ridley championed them here: www.spectator.co.uk/essays/all/5749853/the-global-warming-guerrillas.thtml.

Fiona Fox backed journalism in her blog at: http://fionafox.blogspot.com. Hudson's blogs are here: www.bbc.co.uk/blogs/paulhudson/paul_hudson_/. Littlemore is at www.desmogblog.com/guardian-series-dissects-cru-email-theft-story. The article in the *Columbia Journalism Review* is here http://www.cjr.org/the_observatory/us_press_digs_into_ipcc_story.php?page=all&print=true.

Allen's outburst came here: www.guardian.co.uk/commentisfree/cif-green/2010/feb/04/climate-consensus-under-strain.

Chapter 16: Avalanche at the IPCC

I wrote about IPCC authors' fears of attack in 2006 in *New Scientist*, "State of Denial", 14 November 2006. And reported the Paris meeting in *New Scientist*, "Climate change: what the IPCC didn't tell us", 9 February 2007. The paper by Rahmstorf and Hansen was in *Science*, "A semi-empirical approach to projecting future sea-level rise", v 315, p368.

My original 1999 article quoting Hasnain, "Flooded out", appeared in *New Scientist*, 5 June 1999. My subsequent report on Glaciergate, "Debate heats up over IPCC melting glaciers claim", appeared first online at *New Scientist* on 8 January 2010 at www.newscientist.com/article/dn18363-debate-heats-up-over-ipcc-melting-glaciers-claim.html, and subsequently in the magazine on 16 January as "IPCC in the firing line over glacier claim". Probably the best summary of how the story unfolded is by Steve Connor in the *Independent*, "A Distraction of Himalayan proportions", 23 January 2010: www.independent.co.uk/environment/climate-change/a-distraction-of- himalayan-proportions-1876420.html, with my sidebar here: www.independent.co.uk/opinion/commentators/fred-pearce-i-wrote-the-offending-article-i-stand-by-it-1876419.html.

The WWF Nepal Program report on the Himalayan melt, "An overview of glaciers, glacier retreat, and subsequent impacts in Nepal, India and China" is at http://assets.panda.org/downloads/glacierssummary.pdf. The *Down to Earth* article is archived here: www.indiaenvironmentportal.org.in/node/319. The *Guardian* had Pachauri attacking the Indian government on its glaciers report here: www.guardian.co.uk/environment/2009/nov/09/india-pachauri-climate-glaciers.

The IPCC statement of retraction is here: www.ipcc.ch/pdf/presentations/himalaya-statement-20january2010.pdf.

The Pielke paper on damage from hurricanes is "Hurricane vulnerability in Latin America and the Caribbean: normalised damage and loss potentials", *Natural Hazards Review*, v 4, p101, and the IPCC paragraph is here: www.ipcc.ch/publications_and_data/ar4/wg2/en/ch1s1-3-8-4.html.

Read Agoumi's paper at www.cckn.net/pdf/north_africa.pdf. And the summary to the IPCC Africa chapter here:

www.ipcc.ch/publications_and_data/ar4/wg2/en/ch9s9-es.html.

Arnell's paper is "Climate change and global water resources: SRES emissions and socio-economic scenarios" in *Global Environmental. Change*, v 14, p31.

The launch of the IPCC independent review is here: www.ipcc.ch/press_information/press_information.htm.

Chapter 17: The inquiries

UEA announced the Oxburgh inquiry here: www.uea.ac.uk/mac/comm/media/press/CRUstatements/SAPannounce. It was reviewed by the *Guardian* here: www.guardian.co.uk/environment/2010/mar/22/lord-oxburgh-uea-inquiry, and the *Financial Times* here: http://blogs.ft.com/energy-source/2010/03/22/uea-emails-scientific-panel-announced/.

The Penn State inquiry into Mann reports at www.research.psu.edu/orp/Findings_Mann_Inquiry.pdf.

The *Guardian* reported on the Institute of Physics' submission here: www.guardian.co.uk/environment/2010/mar/05/climate-emails-institute-of-physics-submission. And I reported the Commons oral hearings here: www.guardian.co.uk/environment/ cif-green/2010/mar/01/phil-jones-commons-emails-inquiry. CRU staff are listed here: http://www.cru.uea.ac.uk/cru/people/.

The Commons inquiry's findings and evidence are published at: www.publications.parliament.uk/pa/cm200910/cmselect/cmsctech/387/387i.pdf. The *Guardian* reported it here: www.guardian.co.uk/environment/2010/mar/31/climate-mails-inquiry-jones-cleared, with my thoughts here: www.guardian.co.uk/environment/cif-green/2010/mar/31/hacked-climate-email-inquiry-phil-jones.

Chapter 18: Bonfire of the scientific vanities

Schneider emailed on replication at 1231257056.txt; and Jones to Haughton at 1256765544.txt

Curry's views appear in her interview with *Discover* at http://discovermagazine.com/2010/apr/10-it.s-gettin-hot-in-here-big-battle-over-climate-science, as well as in discussion with Andy Revkin of the *New York Times* here: http://dotearth.blogs.nytimes.com/2009/11/27/a-climate-scientist-on-climate-skeptics/; on ClimateAudit here: http://climateaudit.org/2009/11/22/curry-on-the-credibility-of-climate-research/; and on WUWT here: http://wattsupwiththat.com/2010/02/24/on-the-credibility-of-climate-research-part-ii-towards-rebuilding-trust/.

Mann fights back here: www.climatesciencewatch.org/index.php/csw/details/michael-mann-interview/. Hans von Storch was in the Wall Street Journal here: http://online.wsj.com/article/SB10001424052748704238104574601443947078538.html.

Hulme and Ravetz were on the BBC here: http://news.bbc.co.uk/1/hi/sci/tech/8388485.stm.

RealClimate announced sharing more data here: www.realclimate.org/index.php/data-sources/. And the Met Office did the same here: www.metoffice.gov.uk/corporate/pressoffice/2009/pr20091205.html.

Courrielche diagnoses here: http://bigjournalism.com/pcourrielche/2010/01/08/peer-to-peer-review-how-climategate-marks-the-maturing-of-a-new-science-movement-part-i/.

Chapter 19: Saving the science

The Imperial College research on radiation, "Physics of the Earth's radiative energy balance" by J Harries is in *Contemporary Physics*, v 41, p309. I explore some of the issues raised by sceptics in *New Scientist* in "Greenhouse wars", 19 July 1997. And I reported on Latif's findings in "World 'will cool for next decade' ", *New Scientist*, 9 September 2009. I looked at non-warming trends in the *Guardian* in "Climate of suspicion", www.guardian.co.uk/commentisfree/2008/jun/07/climatechange.

Pielke's remarks on the IPCC were at http://pajamas media.com/blog/pielke-sr-climategate-emails-just-a-small-sample-of-a-broad-issue-pjm-exclusive/.

ACKNOWLEDGMENTS

Numerous climate scientists have helped in making sense of the emails and their context. Some, I fear, may not want to be associated with the outcome. Most are named in my notes on sources, but let me mention some whose contributions are more pervasive. Of the protagonists, Keith Briffa at CRU first told me about what we can and cannot learn from tree rings. Mike Mann, even in the toughest times, has been willing to respond to questions instantly and pithily. Ben Santer, Gavin Schmidt, Steve Schneider, Stefan Rahmstorf and Tom Wigley have all been valuable sources of expertise over the years. More recently, I have corresponded particularly with Steve McIntyre, Doug Keenan, Graham Cogley and Jerry Ravetz.

The original versions of the investigation, as published in the *Guardian* newspaper and online, have been significantly enhanced by the many contributions from protagonists and others in the unique online annotation initiated by the *Guardian*. I especially thank all those who took part in this exercise. Some of the material here was first assembled during research for *New Scientist*, whose editors, as always, deserve thanks for supporting honest and investigative journalism.

At the *Guardian*, George Monbiot's early bravery in response to the release of the emails should be acknowledged. Some of the research into how the emails may have become public was done by David Leigh and Charles Arthur. James Randerson has been a hugely valuable and attentive editor through all stages of the investigation and its appearance in the newspaper, online, through the online annotation exercise and now in book form. I also thank Damian Carrington for his strong backing, Nigel Wilcockson at Random House for his valuable suggestions for the book's structure, and Phil Daoust for his eagle-eyed copy editing. Above all, Ian Katz's enthusiasm and dedication to the project were the vital ingredients that made it happen. The mistakes, of course, are mine.